ECLECTIC EDUCATIONAL SERIES

KEY

TO

RAY'S™ NEW ARITHMETICS

PRIMARY, INTELLECTUAL AND PRACTICAL

Originally published by
Van Antwerp, Bragg & Co.

This edition published by

MOTT MEDIA

1130 Fenway Circle, Fenton, Michigan 48430
www.mottmedia.com

ISBN-10: 0-88062-062-5
ISBN-13: 978-0-88062-062-8
Printed in the United States of America

PRESENT PUBLISHER'S PREFACE

We are honored and happy to bring to you the classic Arithmetics by Joseph Ray. In the 1800s these popular books sold more than any other arithmetics in America, in fact over 120,000,000 copies. Now with this reprinting, they are once again available for America's students.

Ray's Arithmetics are organized in an orderly manner around the discipline of arithmetic itself. They present principles and follow up each one with examples which include difficult problems to challenge the best students. Students who do not master a concept the first time can return to it later, work the more difficult problems, and master the concept. Thus in these compact volumes is a complete arithmetic course to study in school, to help in preparing for ACT and SAT tests, and to use for reference throughout a lifetime.

In order to capture the spirit of the original Ray's, we have refrained from revising the problems and prices. Only a few words have been changed, as we felt it wise. Thus students will have to rely on their arithmetic ability to solve the problems. Also the charm of a former era lives on in this reprinting. Flour and salt are sold by the barrel, kegs may contain tar, and postage stamps cost 3¢ each. Through this content, students learn social history of the 1800s in a unique, hands-on manner at the same time they are mastering arithmetic.

The series consists of four books ranging from Primary Arithmetic to Higher Arithmetic, as well as answer keys to accompany them. We have added a teacher's guide to help today's busy teachers and parents.

We wish to express our appreciation to the staff of the Special Collections Library at Miami University, Oxford, Ohio, for its cooperation in allowing us to use copies of their original Ray's Arithmetics.

George M. Mott, Founder
Mott Media

PREFACE

This Key to Ray's New Arithmetics is presented to the public in the hope that teachers may be assisted thereby in their arduous labors.

It may be supposed that every teacher is able to solve every example in the series; but it should not therefore be required that the work shall actually be done by the teacher. The examination to which every teacher in the Public Schools submits, is the only test that can be required of the possession of knowledge. In the actual work of the school-room, the teacher is clearly entitled to use all appliances that will save him from drudgery, and enable him to devote his full time and energies to the task of imparting instruction and guiding the labors of his pupils.

In most schools, the methods of instruction have so changed that twice as much labor is now expected from the teacher as was required twenty years ago. Much work is now written on slates or paper; and this must be carefully examined, criticized, and marked. This increased labor leaves the teacher little time to work out the examples, and test their correctness.

In the Key to the Intellectual Arithmetic, all the problems which might be supposed to present the least difficulty have been briefly solved. The form of solution is usually the briefest that is consistent with accuracy and clearness; no formal method is pursued.

The Key to the Practical Arithmetic is believed to be complete. It was not thought best to omit anything that might help any teacher. Here, too, the solutions are as brief as possible. Figures and signs are used wherever they serve the purpose. Lengthy explanations are avoided,—the work only being given.

With the hope that this Key will be found to be full and accurate, and so arranged as to prove helpful on all occasions, it is respectfully submitted to the teachers of Ray's Arithmetics.

Cincinnati, 1879.

ANSWERS

to

RAY'S NEW PRIMARY ARITHMETIC

LESSON VI.

(1.) 0; 1, 10; 2, 20; 3, 30; 4, 40; 5, 50; 6, 60; 7, 70; 8, 80; 9, 90.

(2.) 11; 12, 21; 13, 31; 14, 41; 15, 51; 16, 61; 17, 71; 18, 81; 19, 91.

(3.) 22; 23, 32; 24, 42; 25, 52; 26, 62; 27, 72; 28, 82; 29, 92.

(4.) 33; 34, 43; 35, 53; 36, 63; 37, 73; 38, 83; 39, 93.

(5.) 44; 45, 54; 46, 64; 47, 74; 48, 84; 49, 94.

(6.) 55; 56, 65; 57, 75; 58, 85; 59, 95.

(7.) 66; 67, 76; 68, 86; 69, 96.

(8.) 77; 78, 87; 79, 97.

(9.) 88; 89, 98.

(10.) 99; 100.

LESSON VII.

(1.) 1. (2.) 2; 1; 2; 2. (3.) 3; 3; 3. (4.) 2; 1; 3. (5.) 4; 4; 4; 4; 4. (6.) 3; 2; 1; 4; 2. (7.) 5; 5; 5; 5. (8.) 4; 3; 2; 1; 5.

LESSON VIII.

(1.) 6; 6; 6; 6; 6; 6. (2.) 5; 4; 3; 2; 1; 6; 3; 2. (3.) 7; 7; 7; 7; 7. (4.) 6; 5; 4; 3; 2; 1; 7. (5.) 8; 8; 8; 8; 8; 8. (6.) 8; 8; 8. (7.) 7; 6; 5; 4; 3; 2; 1. (8.) 8; 4; 2.

LESSON IX.

(1.) 9; 9; 9; 9; 9; 9; 9. (2.) 9; 9. (3.) 8; 7; 6; 5; 4; 3; 2; 1. (4.) 9; 3. (5.) 10; 10; 10; 10; 10. (6.) 10; 10; 10. (7.) 10; 10; 10. (8.) 9; 8; 7; 6; 5; 4; 3; 2; 1. (9.) 10; 5; 2.

LESSON X.

2; 5; 4; 7. (1.) 7 birds; 11 birds. (2.) 7; 11. (3.) 6 birds; 12 birds. (4.) 6; 12. (5.) 6 flowers.

LESSON XI.

(1.) 3 cents. (2.) 4 raisins. (3.) 5 pears. (4.) 6 cherries. (5.) 7 cents. (6.) 8 plums. (7.) 9 oranges. (8.) 10 peaches. (9.) 11 cents.

LESSON XII.

(1.) 4 birds. (2.) 5 tops. (3.) 6 chestnuts. (4.) 7 apples. (5.) 8 chickens. (6.) 9 cakes. (7.) 10 marbles. (8.) 11 fishes. (9.) 12 cents.

LESSON XIII.

(1.) 5 cents. (2.) 6 dimes. (3.) 7 apples. (4.) 8 pears. (5.) 9 pens. (6.) 10 books. (7.) 11 rabbits. (8.) 12 cents. (9.) 13 cents.

LESSON XIV.

(1.) 6 pens. (2.) 7 pins. (3.) 8 chestnuts. (4.) 9 horses. (5.) 10 cents. (6.) 11 apples. (7.) 12 cents. (8.) 13 dollars. (9.) 14 cents.

LESSON XV.

(1.) 7 chickens. (2.) 8 square blocks. (3.) 9 cents. (4.) 10 cakes. (5.) 11 chairs. (6.) 12 boys. (7.) 13 cents. (8.) 14 letters. (9.) 15 balls.

LESSON XVI.

(1.) 8 cows. (2.) 9 marbles. (3.) 10 cents. (4.) 11 pigs. (5.) 12 plums. (6.) 13 cents. (7.) 14 horses. (8.) 15 miles. (9.) 16 days.

LESSON XVII.

(1.) 9 marbles. (2.) 10 sheep. (3.) 11 boys. (4.) 12 chairs. (5.) 13 apples. (6.) 14 cents. (7.) 13 cents. (8.) 16 roses. (9.) 17 dollars.

LESSON XVIII.

(1.) 10 nuts. (2.) 11 pins. (3.) 12 geese. (4.) 13 marbles. (5.) 14 eggs. (6.) $15. (7.) 16 yards. (8.) 17 years. (9.) 18 cents.

LESSON XIX.

(1.) 11 mice. (2.) 12 cents. (3.) 13 fish. (4.) 14 horses. (5.) 15 oranges. (6.) 16 plums. (7.) 17 buttons. (8.) 18 cents. (9.) 19 pounds.

LESSON XX.

(1.) 12 cents. (2.) $13. (3.) 14 trees. (4.) 15 cows. (5.) 16 yards. (6.) 17 girls. (7.) $18. (8.) 19 nuts. (9.) 20 cents.

LESSON XXI. (Review)

(1.) 6. (2.) 8; 11. (3.) 10, 11. (4.) 11; 10. (5.) 12; 9. (6.) 6; 8. (7.) 9; 11. (8.) 7; 13. (9.) 14; 12. (10.) 13; 12. (11.) 12; 14. (12.) 16; 16. (13.) 14; 12. (14.) 14; 17. (15.) 8; 15. (16.) 13; 19. (17.) 18; 13. (18.) 15; 4. (19.) 10; 10. (20.) 16; 18. (21.) 7; 5. (22.) 15; 17. (23.) 6. (24.) 12; 15. (25.) 16; 16. (26.) 15; 18. (27.) 17; 10. (28.) 17; 14. (29.) 17; 20. (30.) 7; 15.

LESSON XXII. (Review)

(1.) 8; 16. (2.) 13; 17. (3.) 9; 14. (4.) 10; 11. (5.) 10; 11. (6.) 12; 12. (7.) 12; 14. (8.) 13; 19. (9.) 19; 19. (10.) 15; 18. (11.) 11; 12. (12.) 20; 18. (13.) 16; 14. (14.) 13; 12. (15.) 18; 12. (16.) 15; 19. (17.) 13; 19. (18.) 20; 17. (19.) 9; 8. (20.) 20; 20. (21.) 13; 19. (22.) 14; 16. (23.) 14; 15. (24.) 18; 19. (25.) 10; 20. (26.) 14; 13. (27.) 16; 18. (28.) 19; 17. (29.) 9; 11. (30.) 18; 17. (31.) 16. (32.) 17; 14. (33.) 18; 11. (34.) 18; 16. (35.) 20; 15. (36.) 15; 13. (37.) 14; 15. (38.) 19; 15. (39.) 20; 16. (40.) 13; 16. (41.) 1, 3, 5, 7, 9, 11, 13, 15, 17, 19. (42.) 1, 4, 7, 10, 13, 16, 19. (43.) 1, 5, 9, 13, 17. (44.) 1, 6, 11, 16. (45.) 2, 4, 6, 8, 10, 12, 14, 16, 18, 20; 3, 6, 9, 12, 15, 18. (46.) 4, 8, 12, 16, 20; 5, 10, 15, 20. (47.) 2, 5, 8, 11, 14, 17, 20. (48.) 2, 6, 10, 14, 18. (49.) 2, 7, 12, 17. (50) 3, 7, 11, 15, 19.

LESSON XXIII.

(1.) 12 cents. (2.) 14 fish. (3.) 16 cents. (4.) 15 peaches. (5.) $18. (6.) 13 pins. (7.) 17 marbles. (8.) $19. (9.) $19. (10.) 20 cents. (11.) 15 apples; 17 cents. (12.) 10 pears; 15 cents.

LESSON XXIV.

(1.) 1 owl. (2.) 1. (3.) 3 bats. (4.) 3. (5.) 6 leaves. (6.) 6.

LESSON XXV.

(1.) 1 rose. (2.) 3 melons. (3.) 1 nut. (4.) 5 pigeons. (5.) 1 marble. (6.) 7 chickens. (7.) 1 cent. (8.) 1 apple. (9.) 9 plums.

LESSON XXVI.

(1.) 2 apples. (2.) 2 cents. (3.) 3 birds. (4.) 4 mice. (5.) 5 oranges. (6.) 6 marbles. (7.) 7 chickens. (8.) 2 cakes. (9.) 9 yards.

LESSON XXVII.

(1.) 1 orange. (2.) 3 ducks. (3.) 3 persons. (4.) 3 pins. (5.) 5 cents. (6.) 3 chickens. (7.) 7 lemons. (8.) 3. (9.) 9 cents.

LESSON XXVIII.

(1.) 1 apple. (2.) 4 birds. (3.) 3 ducks. (4.) 4 beans. (5.) 4 panes. (6.) 6 trees. (7.) 7 years. (8.) 4 oranges. (9.) 9 chickens.

LESSON XXIX.

(1.) 1 orange. (2.) 5 crows. (3.) 3 ships. (4.) 4 yards. (5.) 5 marbles. (6.) 5 cents. (7.) 7 cents. (8.) $8. (9.) 5 nuts.

LESSON XXX.

(1.) 6 marks. (2.) 2 cents. (3.) 3 raisins. (4.) 6 eggs. (5.) 5 cents. (6.) $6. (7.) 7 dozen. (8.) 6 oranges. (9.) 6; 8.

LESSON XXXI.

(1.) 1 cent. (2.) $7. (3.) 3 oranges. (4.) 4 cents. (5.) 7 marbles. (6.) 6 lemons. (7.) 7 years. (8.) 7 birds. (9.) 8; 7.

LESSON XXXII.

(1.) 8 pecans. (2.) 2 cents. (3.) 3 peaches. (4.) 4 cents. (5.) 8 oranges. (6.) 6 marbles. (7.) 8 pens. (8.) 8 persons. (9.) 8; 7.

LESSON XXXIII.

(1.) 1 walnut. (2.) 2 cents. (3.) $3. (4.) 9 birds. (5.) 5 horses. (6.) 9 plums. (7.) 7 marbles. (8.) 9 horses. (9.) 9 cents.

LESSON XXXIV.

(1.) 1 cent. (2.) $2. (3.) 10 pupils. (4.) 4 trees. (5.) 10 marbles. (6.) 6 cents. (7.) $10. (8.) 8 cents. (9.) 9; 10.

LESSON XXXV. (Review)

(1.) 2. (2.) 6; 4; 3. (3.) 2; 6; 7. (4.) 8; 4; 7. (5.) 8; 10; 6. (6.) 6; 10; 7. (7.) 3; 3; 5. (8.) 9; 9; 5. (9.) 9; 7; 2. (10.) 4; 5; 5. (11.) 2; 3; 6. (12.) 3; 3; 2. (13.) 5; 6; 7. (14.) 8; 9; 8. (15.) 4; 9; 8. (16.) 7; 8; 6. (17.) 4; 6; 5. (18.) 9; 7; 3. (19.) 8; 10; 5. (20.) 7; 9; 10. (21.) 3; 7; 10. (22.) 10; 2; 2. (23.) 4; 8; 3. (24.) 10; 10; 6. (25.) 4; 4; 2. (26.) 8; 9; 9. (27.) 5; 2; 5.

LESSON XXXVI.

(1.) 4. (2.) 2. (3.) 10; 7. (4.) 9; 8. (5.) 2; 3. (6.) 5; 4. (7.) 10; 8. (8.) 2; 2. (9.) 7; 9. (10.) 9; 8. (11.) 2; 5. (12.) 9; 7. (13.) 3; 7. (14.) 10; 10. (15.) 2; 5. (16.) 8; 8. (17.) 6; 9. (18.) 8; 10. (19.) 5; 4. (20.) 6; 4. (21.) 8; 6. (22.) 6; 7. (23.) 4; 7. (24.) 7; 5. (25.) 8; 3. (26.) 3; 7. (27.) 9; 2.

LESSON XXXVII.

(1.) 20, 18, 16, 14, 12, 10, 8, 6, 4, 2, 0. (2.) 18, 15, 12, 9, 6, 3, 0. (3.) 20, 16, 12, 8, 4, 0. (4.) 20, 15, 10, 5, 0. (5.) 18, 12, 6, 0. (6.) 6 cents. (7.) 3. (8.) 2 apples. (9.) 5 cents. (10.) 9 cents. (11.) 5. (12.) 18 oranges. (13.) 5 marbles. (14.) 5 apples.

LESSON XXXVIII.

(1.) 12 birds. (2.) 12. (3.) 16 eggs. (4.) 16. (5.) 16 wings. (6.) 16.

LESSON XXXIX.

(1.) 2 cents. (2.) 3 cents. (3.) 4 cents. (4.) 5 cents. (5.) $6. (6.) 7 miles. (7.) 8 cents. (8.) 9 cents. (9.) 10 cents.

LESSON XL.

(1.) 4 cents. (2.) 6 cents. (3.) 8 cents. (4.) 10 cents. (5.) 12 cents. (6.) 14 cents. (7.) 16 cents. (8.) 18 cents. (9.) 20 cents.

LESSON XLI.

(1.) 6 cents. (2.) 9 cents. (3.) 12 miles. (4.) 15 apples. (5.) 18 plums. (6.) 21 cents. (7.) 24 cents. (8.) 27 cents. (9.) 30 cents.

LESSON XLII.

(1.) 8 feet. (2.) 12 pigeons. (3.) 16 cents. (4.) 20 cents. (5.) 24 quarters. (6.) 28 cents. (7.) 32 cents. (8.) 36 cents. (9.) 40 cents.

LESSON XLIII.

(1.) 10 cents. (2.) 15 miles. (3.) 20 plums. (4.) 25 chickens. (5.) 30 eggs. (6.) 35 cents. (7.) 40 cents. (8.) 45 cents. (9.) 50 cents.

LESSON XLIV.

(1.) 12 yards. (2.) 18 chickens. (3.) 24 panes. (4.) 30 peaches. (5.) 36 cents. (6.) 42 cents. (7.) 48 ounces. (8.) 54 cents. (9.) 60 cents.

LESSON XLV.

(1.) 14 cents. (2.) 21 marbles. (3.) 28 days. (4.) 35 peaches. (5.) 42 miles. (6.) 49 boys. (7.) 56 marbles. (8.) 63 cents. (9.) 70 cents.

LESSON XLVI.

(1.) 16 cents. (2.) 24 fishes. (3.) 32 chickens. (4.) 40 windows. (5.) 48 pints. (6.) 56 cents. (7.) 64 quarts. (8.) 72 cents. (9.) 80 cents.

LESSON XLVII.

(1.) 18 cents. (2.) 27 cents. (3.) 36 panes. (4.) 45 cents. (5.) 54 cents. (6.) 63 cents. (7.) 72 miles. (8.) 81 cents. (9.) 90 cents.

LESSON XLVIII.

(1.) 20 cents. (2.) $30. (3.) 40 pecks. (4.) 50 pounds. (5.) $60. (6.) 70 marbles. (7.) 80 cents. (8.) $90. (9.) $100.

LESSON XLIX. (Review)

(1.) 10. (2.) 12; 18. (3.) 20; 9. (4.) 8; 20. (5.) 16; 4. (6.) 30; 30. (7.) 12; 24. (8.) 40; 21. (9.) 28; 25. (10.) 15; 18. (11.) 36; 24. (12.) 27; 32. (13.) 16; 45. (14.) 35; 80. (15.) 48; 70. (16.) 36; 64. (17.) 63; 54. (18.) 40; 49. (19.) 50; 60. (20.) 42; 90. (21.) 56; 81. (22.) 72; 100. (23.) 6; 14. (24.) 15; 42. (25.) 24; 54. (26.) 40; 63. (27.) 90; 72. (28.) 18. (29.) 16. (30.) 24. (31.) 20. (32.) 30. (33.) 36. (34.) 24. (35.) 42. (36.) 28. (37.) 64. (38.) 36. (39.) 48. (40.) 40. (41.) 60. (42.) 40. (43.) 32. (44.) 48. (45.) 80. (46.) 70. (47.) 90. (48.) 80. (49.) 90. (50.) 60. (51.) 54. (52.) 36. (53.) 45. (54.) 54. (55.) 100. (56.) 72. (57.) 63. (58.) 72.

LESSON L. (Review)

(1.) 9. (2.) 8. (3.) 50. (4.) 42. (5.) 49. (6.) 32. (7.) 27. (8.) 12. (9.) 28. (10.) 25. (11.) 48. (12.) 14. (13.) 64. (14.) 81. (15.) 15. (16.) 45. (17.) 20. (18.) 54. (19.) 18. (20.) 24. (21.) 16. (22.) 21. (23.) 36. (24.) 40. (25.) 56. (26.) 63. (27.) 30.

LESSON LI.

(1.) 4 cents. (2.) $20. (3.) 9 cents. (4.) $14. (5.) $9. (6.) 72 bushels. (7.) $50. (8.) 15 cents. (9.) 28 shoes. (10.) 54 cents.

LESSON LII.

(1.) 5 ducks. (2.) 4. (3.) 2 wings. (4.) 2. (5.) 2 flowers.

LESSON LIII.

(1.) 2 apples. (2.) 3 marbles. (3.) 2 lemons. (4.) 2 peaches. (5.) 6 yards. (6.) 2 oranges. (7.) 8 tops. (8.) 2 kites. (9.) 2 books.

LESSON LIV.

(1.) 2 groups. (2.) 3 yards. (3.) 3 pears. (4.) 5 yards. (5.) 6 oranges. (6.) 3 groups. (7.) 3 yards. (8.) 3 cents. (9.) 10 postage stamps.

LESSON LV.

(1.) 2 oranges. (2.) 3 gallons. (3.) 4 apples. (4.) 5 scholars. (5.) 6 copy-books. (6.) 4 tops. (7.) 8 peaches. (8.) 4 cakes. (9.) 4 books.

LESSON LVI.

(1.) 2 oranges. (2.) 5 pencils. (3.) 5 toy books. (4.) 5 pears. (5.) 5 melons. (6.) 5 weeks. (7.) 5 cakes. (8.) 9 tops. (9.) 10 slates.

LESSON LVII.

(1.) 2 quarts. (2.) 3 oranges. (3.) 4 trees. (4.) 6 pears. (5.) 6 pounds. (6.) 6 lemons. (7.) 6 pencils. (8.) 9 rings. (9.) $10.

LESSON LVIII.

(1.) 2 piles. (2.) 3 pineapples. (3.) 4 melons. (4.) 5 peaches. (5.) 7 miles. (6.) 7 cents. (7.) 7 trees. (8.) 7 yards. (9.) $10.

LESSON LIX.

(1.) 2 pecks. (2.) 3 oranges. (3.) 8 pencils. (4.) 8 cents. (5.) 6 canes. (6.) 8 tops. (7.) 8 peaches. (8.) 8 cents. (9.) 10 cents.

LESSON LX.

(1.) 2 pencils. (2.) 9 cents. (3.) 9 cents. (4.) 9 cents. (5.) 6 cents. (6.) 9 miles. (7.) 8 yards. (8.) 9 blocks. (9.) 9 books.

LESSON LXI.

(1.) 2 melons. (2.) 3 quinces. (3.) 10 pears. (4.) 5 oranges. (5.) 6 dozen. (6.) 7 coats. (7.) 10 kites. (8.) 9 balls. (9.) 10 dimes.

LESSON LXII. (Review)

Column 1: 2, 4, 2, 2, 9, 4, 7, 5, 2, 3, 6, 2, 6, 5, 7, 4.
Column 2: 3, 6, 10, 3, 2, 3, 4, 4, 3, 7, 8, 8, 5, 7, 8, 2.
Column 3: 8, 6, 3, 4, 4, 2, 10, 6, 3, 2, 3, 5, 5, 8, 5, 9.
Column 4: 7, 7, 10, 5, 5, 6, 9, 9, 6, 6, 9, 7, 10, 9, 10, 8.
Column 5: 6, 9, 8, 7, 4, 10, 5, 4, 8, 9, 8, 10, 9, 10, 7, 10.

(1.) 2. (2.) 4. (3.) 4. (4.) 7. (5.) 5. (6.) 5. (7.) 6. (8.) 6. (9.) 7. (10.) 8. (11.) 9. (12.) 9. (13.) 10. (14.) 10. (15.) 6. (16.) 3. (17.) 4. (18.) 10. (19.) 3.

LESSON LXIII.

(1.) 2 cents; 12 cents. (2.) 4 cents; 36 cents. (3.) 30 cents. (4.) 30

cents. (5.) 40 cents. (6.) 49 cents. (7.) 9 cents. (8.) 12 pears. (9.) 4 lemons. (10.) 8 cents. (11.) 3 cents.

LESSON LXIV.

(1.) 12. (2.) [reading across] 13, 14, 15, 16, 17, 18, 19, 20, 21. (3.) 13, 14, 15, 16, 17, 18, 19, 20, 21, 22. (4.) 14, 15, 16, 17, 18, 19, 20, 21, 22, 23. (5.) 15, 16, 17, 18, 19, 20, 21, 22, 23, 24. (6.) 16, 17, 18, 19, 20, 21, 22, 23, 24, 25.

LESSON LXV.

(1.) 17, 18, 19, 20, 21, 22, 23, 24, 25, 26.
(2.) 18, 19, 20, 21, 22, 23, 24, 25, 26, 27.
(3.) 19, 20, 21, 22, 23, 24, 25, 26, 27, 28.
(4.) 20, 21, 22, 23, 24, 25, 26, 27, 28, 29.
(5.) 22, 34, 46, 58, 70, 73, 85, 97, 99.
(6.) 23, 35, 47, 49, 51, 54, 66, 78, 90, 100.
(7.) 24, 36, 48, 60, 72, 75, 87, 99, 100.
(8.) 25, 37, 49, 61, 73, 76, 88, 100.
(9.) 26, 38, 50, 62, 74, 77, 89, 100.
(10.) 27, 39, 51, 63, 75, 78, 90, 100.
(11.) 28, 40, 52, 64, 76, 79, 91, 100.
(12.) 29, 41, 53, 65, 77, 80, 92, 100.
(13.) 30, 42, 54, 66, 78, 81, 93, 100.

LESSON LXVI.

(1.) 31, 43, 55, 67, 79, 82, 94, 96.
(2.) 27, 39, 51, 63, 66, 78, 90, 93.
(3.) 31, 38, 50, 62, 74, 77, 89, 94.
(4.) 32, 44, 49, 61, 73, 85, 88, 91.
(5.) 34, 42, 50, 59, 71, 83, 95, 98.
(6.) 36, 36, 43, 57, 65, 70, 82, 94.
(7.) 37, 37, 48, 56, 64, 75, 81, 93.
(8.) 26, 38, 50, 62, 65, 77, 89, 99.
(9.) 29, 36, 48, 60, 72, 83, 91, 98.
(10.) 40, 55, 64, 73, 82, 91, 98, 98.
(11.) 34, 42, 50, 66, 74, 83, 92, 99.

(12.) 32, 40, 54, 62, 76, 84, 97, 98.
(13.) 35, 47, 56, 64, 77, 83, 95, 97.
(14.) 38, 46, 57, 65, 72, 83, 91, 99.
(15.) 35, 43, 51, 64, 75, 79, 96, 87, 87, 96.

LESSON LXVII.

(1.) 20 cents. (2.) 22 cents. (3.) 25 cents. (4.) $19. (5.) 21 years. (6.) 25 cents. (7.) 23 cents. (8.) 24 cents. (9.) 27 years. (10.) 29 pupils. (11.) 40 chickens. (12.) 21 ducks. (13.) 51 cents. (14.) 27 oranges. (15.) 32 cents. (16.) 50.

LESSON LXVIII.

(1.) 11 marbles. (2.) 13 cents. (3.) 13 plums. (4.) 13 peaches. (5.) 13; 13. (6.) 14 cents. (7.) 29 needles. (8.) 29 chickens. (9.) 29 eggs. (10.) 30 cents. (11.) 49 cents. (12.) 35 quarts. (13.) 47 cents. (14.) 33 marbles. (15.) 56.

LESSON LXIX.

(1.) 13. (2.) 17. (3.) 13. (4.) 18. (5.) 23. (6.) 19. (7.) 14. (8.) 16. (9.) 20. (10.) 27. (11.) 24. (12.) 22. (13.) 21 (14.) 25. (15.) 30. (16.) 29. (17.) 15. (18.) 31. (19.) 32. (20.) 26. (21.) 28. (22.) 33. (23.) 45. (24.) 55. (25.) 65. (26.) 56. (27.) 80. (28.) 83. (29.) 89.

LESSON LXX.

(1.) 14 cents. (2.) 15 cents. (3.) 10 cents. (4.) 10 oranges were given away; 12 oranges were left. (5.) 30 cents. (6.) 5 cents. (7.) 12 cents. (8.) 28 years. (9.) 65 cents. (10.) 4 cents. (11.) 40 cents spent; 10 cents left.

LESSON LXXI.

(1.) 7. (2.) 29; 41. (3.) 71; 11. (4.) 14; 4. (5.) 30; 16. (6.) 44; 57. (7.) 43; 31. (8.) 39; 27. (9.) 15; 53. (10.) 22; 67. (11.) 2; 21. (12.) 7; 17. (13.) 0; 44. (14.) 68; 7. (15.) 21. (16.) 87. (17.) 57. (18.) 23. (19.) 25. (20.) 36. (21.) 37. (22.) 33. (23.) 20. (24.) 40. (25.) 53. (26.) 10. (27.) 10. (28.) 97.

LESSON LXXII.

(1.) 2. (2.) 3; 2. (3.) 4; 2. (4.) 3. (5.) 5; 2. (6.) 6; 4; 3; 2. (7.) 7; 2. (8.) 5; 3. (9.) 8; 4; 2. (10.) 9; 6; 3; 2. (11.) 10; 5; 4; 2. (12.) 7; 3. (13.) 8; 6; 4; 3. (14.) 5. (15.) 9; 3. (16.) 7; 4. (17.) 10; 6; 5; 3. (18.) 8; 4.

LESSON LXXIII.

(1.) 7; 5. (2.) 9; 6; 4. (3.) 10; 8; 5; 4. (4.) 7; 6. (5.) 9; 5. (6.) 8; 6. (7.) 7. (8.) 10; 5. (9.) 9; 6. (10.) 8; 7. (11.) 10; 6. (12.) 9; 7. (13.) 8. (14.) 10; 7. (15.) 9; 8. (16.) 10; 8. (17.) 9. (18.) 10; 9. (19.) 10.

LESSON LXXIV.

(1.) 16 cents. (2.) 16 oranges. (3.) Jane is 14 years old; 20 years. (4.) 3; 5; 7; 8. (5.) 5 lemons. (6.) 15 men. (7.) 10 miles; 2 hours. (8.) 37 cents. (9.) 36 days. (10.) 6 oranges; 8 oranges; 10 oranges. (11.) 44 cents. (12.) $56.

LESSON LXXV.

(1.) Charles—4 marbles; Henry—16 marbles. (2.) 3 cents. (3.) 23 cents. (4.) 10 cents. (5.) 9 cents; 3 cents. (6.) 6 cents. (7.) 7 flags; 49 cents. (8.) 45 cents. (9.) 35 cents. (10.) 50 cents. (11.) 10 cents. (12.) 48 cents. (13.) $20.

LESSON LXXVI.

(1.) 2 oranges. (2.) 12 cents; 6 coins. (3.) 4 oranges. (4.) 6 tops. (5.) 2 quarts. (6.) 6 barrels. (7.) 5 quarts. (8.) 8 barrels. (9.) 4 melons. (10.) 5 oranges. (11.) 5 pears.

LESSON LXXVII.

(1.) 3 pineapples. (2.) 8 pencils. (3.) 9 kites. (4.) 8 peaches. (5.) 5 quarts. (6.) 10 pears. (7.) 2 days. (8.) 6 yards. (9.) 4 slates. (10.) 5 bottles. (11.) 5 oranges. (12) 12 days. (13.) $60. (14.) 12 trees.

LESSON LXXVIII.

(1.) 2, 4, 6, 8, 10, 12, 14, 16, 18, 20, 22, 24, 26, 28, 30, 32, 34,

36, 38, 40, 42, 44, 46, 48, 50, 52, 54, 56, 58, 60, 62, 64, 66, 68, 70, 72, 74, 76, 78, 80, 82, 84, 86, 88, 90, 92, 94, 96, 98, 100. (2.) Same numbers as in one but in reverse order and the addition of 0. (3.) 3, 6, 9, 12, 15, 18, 21, 24, 27, 30, 33, 36, 39, 42, 45, 48, 51, 54, 57, 60, 63, 66, 69, 72, 75, 78, 81, 84, 87, 90, 93, 96, 99. (4.) See three plus 0. (5.) 4, 8, 12, 16, 20, 24, 28, 32, 36, 40, 44, 48, 52, 56, 60, 64, 68, 72, 76, 80, 84, 88, 92, 96, 100. (6.) See five plus 0. (7.) 5, 10, 15, 20, 25, 30, 35, 40, 45, 50, 55, 60, 65, 70, 75, 80, 85, 90, 95, 100. (8.) See seven plus 0. (9.) 6, 12, 18, 24, 30, 36, 42, 48, 54, 60, 66, 72, 78, 84, 90, 96. (10.) See nine plus 0. (11.) 7, 14, 21, 28, 35, 42, 49, 56, 63, 70, 77, 84, 91, 98. (12.) See eleven plus 0. (13.) 8, 16, 24, 32, 40, 48, 56, 64, 72, 80, 88, 96. (14.) See thirteen plus 0. (15.) 9, 18, 27, 26, 45, 54, 63, 72, 81, 90, 99. (16.) See fifteen plus 0. (17.) 1, 3, 5, 7, 9, 11, 13, 15, 17, 19, 21, 23, 25, 27, 29 31, 33, 35, 37, 39, 41, 43, 45, 47, 49, 51, 53, 55, 57, 59, 61, 63, 65, 67, 69, 71, 73, 75, 77, 79, 81, 83, 85, 87, 89, 91, 93, 95, 97, 99. (18.) 100, 97, 94, 91, 88, 85, 82, 79, 76, 73, 70, 67, 64, 61, 58, 55, 52, 49, 46, 43, 40, 37, 34, 31, 28, 25, 22, 19, 16, 13, 10, 7, 4, 1. (19.) See eighteen. (20.) 97, 93, 89, 85, 81, 77, 73, 69, 65, 61, 57, 53, 49, 45, 41, 37, 33, 29, 25, 21, 17, 13, 9, 5, 1. (21.) See twenty. (22.) 96, 91, 86, 81, 76, 71, 66, 61, 56, 51, 46, 41, 36, 31, 26, 21, 16, 11, 6, 1. (23.) See twenty-two. (24.) 97, 91, 85, 79, 73, 67, 61, 55, 49, 43, 37, 31, 25, 19, 13, 7, 1. (25.) See twenty-four. (26.) 100, 94, 88, 82, 76, 70, 64, 58, 52, 46, 40, 34, 28, 22, 16, 10, 4. (27.) 1, 8, 15, 22, 29, 36, 43, 50, 57, 64, 71, 78, 85, 92, 99. (28.) 100, 93, 86, 79, 72, 65, 58, 51, 44, 37, 30, 23, 16, 9, 2. (29.) 1, 9, 17, 25, 33, 41, 49, 57, 65, 73, 81, 89, 97. (30.) 100, 92, 84, 76, 68, 60, 52, 44, 36, 28, 20, 12, 4. (31.) 1, 10, 19, 28, 37, 46, 55, 64, 73, 82, 91, 100.

LESSON LXXIX.

(1.) 30 cents. (2.) 30 dimes; 70 dimes. (3.) 6 dimes. (4) $60. (5.) $4. (6.) 100 dimes; 80 dimes. (7.) 3 dimes. (8.) 4 dimes. (9.) $5. (10.) 90 cents; 9 dimes. (11.) 15 cents.

LESSON LXXX.

(1.) 12 farthings. (2.) 5 shillings. (3.) 72 pence; 96 pence; 84 pence. (4.) 40 shillings; 80 shillings; 100 shillings. (5.) 2 pounds; 3 pounds; 5 pounds. (6.) 2 pounds. (7.) 15 pence. (8.) 10 books. (9.) 8 tops. (10.) 2 shillings. (11.) 1 pound. (12.) 13 shillings.

LESSON LXXXI.

(1.) 60 ounces. (2.) 8 pounds. (3.) 60 penny-weights; 100 penny-weights. (4.) 3 ounces; 4 ounces. (5.) 48 grains. (6.) 2 grains. (7.) 60 dimes. (8.) 12 ounces. (9.) 36 pounds. (10.) 10 penny-weights. (11.) 36 ounces. (12.) 1 ounce.

LESSON LXXXII.

(1.) 32 ounces; 48 ounces. (2.) 2 pounds; 4 pounds. (3.) 64 ounces; 80 ounces. (4.) 3 pounds; 5 pounds. (5.) 4000 pounds; 6000 pounds. (6.) 2 tons; 3 tons. (7.) 2 pounds. (8.) 64 ounces. (9.) 6 tons. (10.) 5 pounds. (11.) 48 cents; 28 cents. (12.) $8. (13.) 600 pounds. (14.) 10,000 pounds (15.) 80 ounces.

LESSON LXXXIII.

(1.) 4 pints; 6 pints; 8 pints. (2.) 6 quarts; 7 quarts; 8 quarts. (3.) 8 pecks; 12 pecks, 20 pecks. (4.) 4 bushels; 6 bushels; 8 bushels. (5.) 32 pints; 48 pints. (6.) 64 quarts; 96 quarts. (7.) 28 cents. (8.) 1 cent. (9.) 3 pecks. (10.) 16 quarts. (11.) $12; $6. (12.) 2 pecks; 70 cents. (13.) 7 bushels. (14.) 9 pecks.

LESSON LXXXIV.

(1.) 16 gills; 24 gills; 32 gills. (2.) 5 pints; 7 pints; 9 pints. (3.) 12 pints; 16 pints; 20 pints. (4.) 6 quarts; 8 quarts; 9 quarts. (5.) 20 quarts; 28 quarts; 36 quarts. (6.) 4 gallons; 6 gallons; 8 gallons. (7.) 40 cents. (8.) 3 cents. (9.) 10 cents. (10.) 64 cents. (11.) 6 gallons. (12.) $8. (13.) 10 cents. (14.) 15 quarts. (15.) 40 pints.

LESSON LXXXV.

(1.) 36 inches; 60 inches; 84 inches. (2.) 2 feet; 4 feet; 8 feet. (3.) 12 feet; 24 feet; 30 feet. (4.) 3 yards; 5 yards; 9 yards.

(5.) 640 rods; 960 rods. (6.) 12 inches. (7.) 18 yards. (8.) 60 feet. (9.) 2 yards. (10.) 5 feet. (11.) 9 bricks. (12.) 8 feet. (13.) 3 miles. (14.) 360 inches.

LESSON LXXXVI.

(1.) 27 sq. ft.; 45 sq. ft. (2.) 4 sq. yds.; 8 sq. yds. (3.) 320 sq. rds.; 480 sq. rds. (4.) 2 acres; 3 acres. (5.) 27 sq. ft. (6.) 20 sq. yds. (7.) 9 sq. yds. (8.) 20 yards. (9.) 42 cents. (10.) 5 sq. ft. (11.) 36 sq. ft. (12.) 32 sq. ft. (13.) 2 sq. miles (14.) 90 sq. ft.

LESSON LXXXIX.

(1.) 50 cents; 30 cents; 70 cents. (2.) 10 pence; 8 pence. (3.) 3 eagles; 5 eagles; 7 eagles. (4.) 60 shillings; 100 shillings; 80 shillings. (5.) 80 penny-weights; 40 penny-weights. (6.) 3 pounds Troy; 5 pounds Troy. (7.) 32 ounces; 48 ounces. (8.) 2 pounds; 3 pounds. (9.) 48 quarts; 64 quarts; 80 quarts. (1.) 5 bushels; 7 bushels; 9 bushels. (11.) 28 quarts; 24 quarts; 32 quarts. (12.) 4 gallons; 5 gallons; 9 gallons. (13.) 21 feet; 30 feet; 36 feet. (14.) 5 sq. yds.; 7 sq. yds. (15.) 27 scruples; 24 scruples; 30 scruples. (16.) 54 cu. ft. (17.) 2 cu. yds. (18.) 48 hours; 72 hours. (19.) 2 days; 3 days. (20.) 28 weeks; 48 weeks. (21.) 120 minutes; 180 minutes. (22.) 80 quires; 60 quires. (23.) 144 pens. (24.) 42 feet; 30 feet; 54 feet. (25.) 5 feet. (26.) 6 chains; 7 chains; 9 chains. (27.) 1920 farthings; 9600 farthings. (28.) 4 ounces. (29.) 48 sheets; 960 sheets. (30.) 70.

ANSWERS
to
RAY'S INTELLECTUAL ARITHMETIC

LESSON I.

(1.) 2 apples. (2.) 3¢. (3.) 4 marbles. (4.) 5¢. (5.) 4 cakes. (6.) 5 oranges. (7.) 6 apples. (8.) 6 plums. (9.) 6¢. (10.) 7 pears. (11.) 5 fingers; 10 fingers. (12.) 8¢. (13.) 8 cakes. (14.) 9¢. (15.) 8 oranges. (16.) 10¢. (17.) 10 boys. (18.) $9. (19.) 10 peaches. (20.) $7. (21.) 11¢. (22.) 10 pounds. (23.) 10 oranges. (24.) 7¢. (25.) 10. (26.) 8¢.

LESSON II.

(1.) 2; 3; 4; 5; 4; 6; 7; 7; 8; 9; 10; 9. (2.) 7; 11; 9. (3.) 7; 10; 9; 5; 8; 9. (4.) 11; 8; 9; 10; 11; 11; 12. (5.) 10; 11; 13; 11; 12; 11; 13; 12; 13. (6.) 12; 12; 12; 13; 11; 13; 12; 12; 13; 17. (7.) 10; 13; 13; 8; 12; 13; 14; 14; 6+4. (8.) 4; 6; 8; 10; 12; 14; 16. (9.) 1; 4; 7; 10; 13. (10.) 3+9; 2+3+7. (11.) 14. (12.) 14. (13.) 11. (14.) 14. (15.) 15 dozen. (16.) 10¢. (17.) 10. (18.) 13; 13. (19.) 10¢, (20.) $11. (21.) 11¢. (22.) 16 yards. (23.) 18¢. (24.) 14; 18; 20. (25.) 21¢.

LESSON IV.

(1.) 11; 15. (2.) 8; 15; 14; 16. (3.) 14; 17; 15; 13; 16; 12; 14. (4.) 14; 17; 15; 19; 16; 18; 16. (5.) 20; 18; 21; 19; 17. (6.) 16; 18; 20; 22; 21. (7.) 13; 15; 17; 19; 14; 22. (8.) 15; 16; 18; 20; 23; 24. (9.) 17; 19; 18; 20; 22; 23; 21; 24; 25. (10.) 18; 20; 22; 19; 21; 24; 23; 25; 26. (11.) 20; 22; 24; 19; 23; 25; 27; 26. (12.) 20; 22; 24; 21; 23; 25; 27; 26; 28. (13.) 23; 21; 24; 22; 26; 25; 27; 29; 28. (14.) 28; 22; 25; 23; 26; 24; 27; 29; 30. (15.) 24; 22; 21; 26; 28; 27; 29; 25; 31; 30. (16.) 31; 51; 71; 41; 61; 81; 101. (17.) 32; 52; 62; 42; 72; 82; 92; 102. (18.) 36; 56; 46; 66; 86; 76; 96; 106. (19.) 37; 57; 47; 77; 67; 87. (20.) 28; 38; 58; 78; 68; 88; 98; 108. (21.) 31; 51; 71; 91; 101. (22.) 35; 45; 55; 75; 65; 95. (23.) 24; 34; 54;

64; 44; 74; 94; 84; 104. (24.) 37; 57; 47; 67. (25.) 21; 31; 31; 30; 30; 42; 54; 51; 63; 65; 72; 81; 92. (26.) 20; 31; 32; 41; 40; 51; 50; 60; 61; 70; 90; 100; 91. (27.) 33; 34; 43; 45; 53; 56; 63; 65; 73; 83; 81; 93; 95; 103. (28.) 27; 36; 56; 45; 65; 66; 76; 79; 86; 96; 106. (29.) 27; 38; 36; 34; 42; 46; 48; 70; 57; 59; 63; 74.

LESSON V.

(1.) 13. (2.) 16. (3.) 13. (4.) 15. (5.) 17. (6.) 17. (7.) 18. (8.) 19. (9.) 21. (10.) 32. (11.) 34. (12.) 33. (13.) 39. (14.) 39. (15.) 51. (16.) 52. (17.) 50. (18.) 56. (19.) 57. (20.) 53. (21.) 71. (22.) 82. (23.) 43. (24.) 51. (25.) 78. (26.) 15¢. (27.) $15. (28.) 23 books. (29.) $23. (30.) 26 marbles. (31.) 28¢. (32.) $23. (33.) 33¢. (34.) 41 hogs. (35.) 32¢. (36.) 34¢. (37.) 26 years. (38.) 22 marbles. (39.) 2, 4, 6, 8, 10, 12, 14, 16, 18, 20 . . . 90, 92, 94, 96, 98, 100. (40.) 3, 6, 9, 12, 15, 18, 21 . . . 78, 81, 84, 87, 90, 93, 96, 99. (41.) 4, 8, 12, 16, 20, 24, 28, 32, 36, 40, 44, 48, 52, 56, 60. 64, 68, 72, 76, 80, 84, 88, 92, 96, 100. (42.) 5, 10, 15, 20, 25, 30, 35, 40, 45, 50, 55, 60, 65, 70, 75, 80, 85, 90, 95, 100. (43.) 6, 12, 18, 24, 30, 36, 42, 48, 54, 60, 66, 72, 78, 84, 90, 96, 102. (44.) 7, 14, 21, 28, 35, 42, 49, 56, 63, 70, 77, 84, 91, 98. (45.) 8, 16, 24, 32, 40, 48, 56, 64, 72, 80, 88, 96, 104. (46.) 9, 18, 27, 36, 45, 54, 63, 72, 81, 90, 99. (47.) 1, 4, 7, 10, 13, 16, 19, 22, 25, 28, 31, 34, 37, 40, 43, 46, 49, 52, 55, 58, 61, 64, 67, 70, 73, 76, 79, 82, 85, 88, 91, 94, 97, 100. (48.) 3, 7, 11, 15, 19, 23, 27, 31, 35, 39, 43, 47, 51, 55, 59, 63, 67, 71, 75, 79, 83, 87, 91, 95, 99, 103. (49.) 2, 7, 12, 17 22, 27, 32, 37, 42, 47, 52, 57, 62, 67, 72, 77, 82, 87, 92, 97, 102. (50.) 5, 11, 17, 23, 29, 35, 41, 47, 53, 59, 65, 71, 77, 83, 89, 95, 101, 107. (51.) 6, 13, 20, 27, 34, 41, 48, 55, 62, 69, 76, 83, 90, 97, 104. (52.) 7, 15, 23, 31, 39, 47, 55, 63, 71, 79, 87, 95, 103. (53.) 8, 17, 26, 35, 44, 53, 62, 71, 80, 89, 98, 107.

LESSON VI.

(1.) 1 apple. (2.) 2 apples. (3.) 3¢. (4.) 4; 5; 6; 7; 8; 9. (5.) 2¢. (6.) 3 apples. (7.) 4; 5; 6; 7; 8; 9. (8.) 2¢. (9.) 3; 4; 5; 6; 7; 8; 9. (10.) 5 marbles. (11.) 6; 7; 8; 9; 10; 11. (12.) 5 apples. (13.) 6; 7; 8; 9; 10; 11. (14.) 5 marbles. (15.) 6; 7; 8; 9; 10; 11. (16.) 5¢.

(17.) 6; 7; 8; 9; 10; 11; 12. (18.) 5 apples. (19.) 6; 7; 8; 9; 10; 11; 12. (20.) 4 apples. (21.) 5; 6; 7; 8; 9; 10; 11. (22.) 12¢. (23.) 4 years. (24.) $11. (25.) 8 marbles.

LESSON VIII.

(1.) 2¢. (2.) $4. (3.) 5¢. (4.) 9¢. (5.) 6¢. (6.) 4 nuts. (7.) 6 miles. (8.) 8¢. (9.) 5 marbles. (10.) $8. (11.) $8. (12.) 9¢; 8¢. (13.) 2 apples. (14.) 4¢. (15.) 20 marbles. (16.) 50 barrels. (17.) 13 bushels. (18.) 24 years. (19.) 35¢. (20.) 18. (21.) 28. (22.) 10. (23.) 20. (24.) 7. (25.) 13.

LESSON IX.

(1.) 8 marbles. (2.) 10 barrels. (3.) 11¢. (4.) $10. (5.) $9. (6.) $11. (7.) 7¢. (8.) 11 marbles. (9.) 23 marbles; 9 marbles. (10.) 16 marbles. (11.) 4 sheep. (12.) $7. (13.) 15¢. (14.) $9. (15.) 15.

LESSON X.

(1.) 2. (2.) 3. (3.) 5. (4.) 7. (5.) 9. (6.) 12. (7.) 2. (8.) 2. (9.) 5. (10.) 9. (11.) 7. (12.) 6. (13.) 8. (14.) 11. (15.) 10. (16.) 6. (17.) 5. (18.) 5. (19.) 5. (20.) 4. (21.) 5. (22.) 8. (23.) 12. (24.) 12. (25.) 10. (26.) 1. (27.) 11.

LESSON XI.

(1.) 9¢. (2.) 8 gallons. (3.) 11. (4.) 22. (5.) $15. (6.) $17. (7.) 2. (8.) $15. (9.) 16 (10.) $5. (11.) 10 sheep. (12.) $3. (13.) 6. (14.) 5. (15.) 38. (16.) $20. (17.) $80. (18.) 68. (19.) $40. (20.) 8 gallons. (21.) 8. (22.) 1.

LESSON XII.

(1.) 4¢. (2.) 6¢. (3.) 8¢. (4.) 9¢. (5.) 12¢. (6.) 15¢. (7.) 16¢. (8.) $20. (9.) $24. (10.) 25¢. (11.) 30¢. (12.) 36¢. (13.) 14¢. (14.) 21¢. (15.) 28¢. (16.) 35 apples. (17.) 16 apples. (18.) 24¢; 32¢. (19.) 40 apples. (20.) 18¢; 27¢. (21.) 40¢. (22.) 33¢.

LESSON XIII.

(1) 28. (2.) 16; 35; 42. (3.) 48; 54; 63. (4.) 72; 49; 56. (5.) 24;

20; 18. (6.) 50; 44; 36. (7.) 64; 66; 18. (8.) 70; 48; 40. (9.) 6; 8; 10. (10.) 12; 14; 15. (11.) 20; 36; 22. (12.) 24; 25; 33. (13.) 60; 72; 30. (14.) 108; 36; 40. (15.) 110; 120; 96. (16.) 45; 81; 121. (17.) 132; 55; 60. (18.) 80; 84; 88. (19.) 99; 100; 132.

LESSON XIV.

(1.) 14¢. (2.) 21¢. (3.) 30¢. (4.) 56¢. (5.) $48. (6.) $27. (7.) $28. (8.) 80¢. (9.) $42. (10.) $40. (11.) 56 miles. (12.) 64 squares. (13.) 77 trees. (14.) $54. (15.) 72¢. (16.) 96¢. (17.) 18 miles. (18.) 6 days. (19.) 12 days. (20.) 24 days. (21.) 32 days. (22.) $48. (23.) 63 men. (24.) 88 persons. (25.) $56.

LESSON XV.

(1.) 15¢. (2.) 20 miles. (3.) 144 miles. (4.) $12; $3. (5.) 39¢. (6.) 4. (7.) 56 shillings. (8.) $38. (9.) $56. (10.) 36 dimes or $3.60. (11.) 120 horses. (12.) 24. (13.) 60. (14.) 40. (15.) 20¢. (16.) $1. (17.) Albert has 40 marbles; Edward has 60 marbles; 20. (18.) $8. (19.) $10. (20.) 21 miles. (21.) 48 miles. (22.) nothing. (23.) 16¢. (24.) 18¢. (25.) $1.16. (26.) 5¢. (27.) $1. (28.) $12; $32. (29.) $4. (30.) 42 books. (31.) $1.08. (32.) $52. (33.) 19¢. (34.) 63. (35.) 24. (36.) 144. (37.) $77. (38.) 64.

LESSON XVI.

(1.) 4 cakes. (2.) 2 apples. (3.) 3 boys. (4.) 4 apples (5.) 2 peaches. (6.) 3 pears. (7.) 5 cakes. (8.) 7 balls. (9.) 3 lemons. (10.) 8 piles. (11.) 6 plums. (12.) 4 oranges. (13.) 7 yards. (14.) 11 yards. (15.) 4 oranges; 3 oranges. (16.) 5 trees. (17.) 9 hours. (18.) 7 sheep. (19.) 10 marbles. (20.) 8 dimes. (21.) 7 trees. (22.) $6. (23.) 9 gallons. (24.) 4 hours. (25.) 7¢. (26.) 5 oranges.

LESSON XVIII.

(1.) 6. (2.) 8; 12; 3; 5; 7; 9; 2; 5; 7; 9; 12. (3.) 3; 6; 9; 12; 3; 4; 6; 7; 9; 11. (4.) 2; 4; 6; 8; 9; 12; 3; 5; 7; 9; 12. (5.) 2; 3; 5; 6; 7; 9; 12; 2; 6; 9; 10. (6.) 5; 7; 9; 10; 11; 2; 4; 5; 6; 8; 9; 10; 12. (7.) 4 peaches. (8.) 6 apples. (9.) 10¢. (10.) 7¢. (11.) 8 weeks. (12.) 5 pears; 7 pears; 9 pears. (13.) 5 oranges; 7 oranges; 9 oranges;

11 oranges. (14.) 6 days. (15.) 8 days. (16.) 7 hours. (17.) 3 tons; 5 tons; 6 tons; 7 tons. (18.) $6. (19.) 8¢. (20.) $9. (21.) 12 trees. (22.) $12; $2. (23.) $2. (24.) $12; $4. (25.) 5 miles.

LESSON XIX.

(1.) 6. (2.) 8; 4; 3; 2. (3.) 6; 9; 12; 8. (4.) 4 oranges. (5.) 12 revolutions. (6.) 12 trees. (7.) 30 marbles. (8.) 7 barrels. (9.) 20 hours. (10.) 4 times. (11.) 6. (12.) 36. (13.) 7¢. (14.) 6 men. (15.) 5. (16.) 34 dimes. (17.) 3. (18.) 3¢. (19.) 9. (20.) 18 apples. (21.) The first number, 10. (22.) 30 apples. (23.) 63 miles. (24.) 40¢. (25.) 45 lessons. (26.) $56. (27.) 4. (28.) 6 barrels. (29.) 49. (30.) 9 yards. (31.) $21. (32.) 11 pounds. (33.) 10 hours. (34.) 6 men. (35.) 12 men.

LESSON XX.

(1.) 4; 6; 8; 10; 12; 14; 16; 18; 20. (2.) 6; 9; 12; 15; 18; 21; 24; 27; 30. (3.) 8; 12; 16; 20; 24; 28; 32; 36; 40. (4.) 10; 15; 20; 25; 30; 35; 40; 45; 50. (5.) 12; 18; 24; 30; 36; 42; 48; 54; 60. (6.) 14; 21; 28; 35; 42; 49; 56; 63; 70. (7.) 16; 24; 32; 40; 48; 56; 64; 72; 80. (8.) 18; 27; 36; 45; 54; 63; 72; 81; 90. (9.) 9; 15; 21; 27. (10.) 20; 40; 15. (11.) 66; 22; 55; 99.

LESSON XXII.

(1.) 1¢. (2.) 1¢. (3.) $4. (4.) 2 apples. (5.) 6¢. (6.) $4; $6; $8. (7.) 10¢. (8.) 6. (9.) 15. (10.) 6; 12; 20. (11.) 4; 9; 16; 25; 36. (12.) 6; 15; 28. (13.) 2; 8; 15; 28; 40. (14.) 3; 14; 27. (15.) 2; 6; 12; 20; 30; 42; 56. (16.) 10; 21; 44. (17.) 2¢. (18.) $3. (19.) 10¢. (20.) $18. (21.) 50¢. (22.) 45¢. (23.) $70.

LESSON XXIII.

(1.) $\frac{1}{2}$ apple. (2.) $\frac{1}{3}$ pear. (3.) $\frac{2}{3}$ yard. (4.) $\frac{3}{4}$ orange. (5.) $\frac{2}{5}$; $\frac{3}{5}$; $\frac{4}{5}$. (6.) $\frac{5}{6}$. (7.) $\frac{4}{7}$. (8.) $\frac{7}{10}$. (9.) $\frac{5}{8}$. (10.) $\frac{5}{7}$. (11.) $\frac{3}{8}$; $\frac{3}{10}$; $\frac{3}{11}$; $\frac{3}{20}$. (12.) $\frac{4}{9}$; $\frac{4}{11}$; $\frac{4}{15}$; $\frac{4}{25}$. (13.) $\frac{5}{8}$; $\frac{5}{9}$; $\frac{5}{12}$; $\frac{5}{16}$. (14.) $\frac{2}{15}$; $\frac{7}{15}$; $\frac{8}{15}$; $\frac{11}{15}$; $\frac{13}{15}$. (15.) $\frac{3}{20}$; $\frac{7}{20}$; $\frac{11}{20}$; $\frac{13}{20}$; $\frac{17}{20}$. (16.) $\frac{18}{23}$. (17.) $\frac{24}{35}$. (18.) $\frac{14}{19}$. (19.) $\frac{2}{3}$. (20.) $\frac{9}{20}$. (21.) $\frac{4}{9}$. (22.) $\frac{6}{25}$. (23.) $\frac{8}{21}$. (24.) $\frac{27}{35}$.

LESSON XXIV.

(1.) 2¢. (2.) 6¢. (3.) 9¢. (4.) $12. (5.) 25¢. (6.) 15¢. (7.) 42¢. (8.) 21. (9.) 27; 33. (10.) 16; 20; 28; 32; 40; 44. (11.) 25; 35; 45; 55. (12.) 36; 42; 48; 54; 66; 72. (13.) 49; 77. (14.) 64; 72; 80. (15.) 81; 99. (16.) 100. (17.) 9¢. (18.) 16 years. (19.) 25. (20.) 32. (21.) 35. (22). 20. (23.) 55. (24.) 77. (25.) 36.

LESSON XXV.

(1.) $1\frac{1}{2}$ apples. (2.) $1\frac{1}{3}$ oranges. (3.) $2\frac{1}{2}$¢. (4.) $1\frac{2}{3}$ dollars. (5.) $1\frac{1}{4}$¢. (6.) $1\frac{1}{5}$. (7.) $3\frac{1}{2}$; $2\frac{1}{3}$; $1\frac{3}{4}$; $1\frac{2}{5}$; $1\frac{1}{6}$. (8.) $2\frac{2}{3}$; $1\frac{3}{5}$; $1\frac{1}{7}$. (9.) $4\frac{1}{2}$; $2\frac{1}{4}$; $1\frac{4}{5}$; $1\frac{2}{7}$; $1\frac{1}{8}$. (10.) $3\frac{1}{3}$; $1\frac{3}{7}$; $1\frac{1}{9}$. (11.) $5\frac{1}{2}$; $3\frac{2}{3}$; $2\frac{3}{4}$; $2\frac{1}{5}$; $1\frac{5}{6}$; $1\frac{4}{7}$; $1\frac{3}{8}$; $1\frac{2}{9}$; $1\frac{1}{10}$. (12.) $2\frac{2}{5}$; $1\frac{5}{7}$; $1\frac{1}{11}$. (13.) $2\frac{1}{2}$ apples. (14.) $2\frac{1}{3}$ yards. (15.) $3\frac{1}{4}$ yards. (16.) $3\frac{2}{5}$ pints. (17.) $3\frac{5}{6}$ barrels. (18.) $3\frac{4}{7}$ yards. (19.) $6\frac{1}{2}$. (20.) $7\frac{1}{2}$; $8\frac{1}{2}$; $9\frac{1}{2}$; $10\frac{1}{2}$; $11\frac{1}{2}$. (21.) $6\frac{2}{3}$; $8\frac{2}{3}$; $9\frac{2}{3}$; $10\frac{1}{3}$; $11\frac{2}{3}$. (22.) $6\frac{3}{4}$; $8\frac{1}{4}$; $9\frac{3}{4}$; $10\frac{1}{4}$; $11\frac{3}{4}$. (23.) $7\frac{1}{6}$; $7\frac{5}{6}$; $8\frac{1}{6}$; $9\frac{5}{6}$; $10\frac{1}{6}$. (24.) $3\frac{3}{7}$; $4\frac{4}{7}$; $5\frac{5}{7}$; $6\frac{6}{7}$; $8\frac{1}{7}$. (25.) $5\frac{5}{8}$; $6\frac{7}{8}$; $8\frac{3}{8}$; $8\frac{7}{8}$; $10\frac{1}{8}$. (26.) $3\frac{7}{9}$; $4\frac{2}{9}$; $5\frac{5}{9}$; $6\frac{4}{9}$; $7\frac{1}{9}$. (27.) $6\frac{3}{10}$; $6\frac{9}{10}$; $7\frac{7}{10}$; $8\frac{3}{10}$; $9\frac{1}{10}$. (28.) $3\frac{9}{11}$; $4\frac{2}{11}$; $4\frac{10}{11}$; $5\frac{5}{11}$; $6\frac{4}{11}$. (29.) $5\frac{2}{3}$; $6\frac{7}{12}$; $7\frac{1}{12}$; $7\frac{5}{12}$; $7\frac{11}{12}$.

LESSON XXVI.

(1.) 5. (2.) 7; 9; 11; 13. (3.) 13; 17; 19; 23. (4.) 13; 19; 21; 27. (5.) 6; 17; 32; 36. (6.) 17; 29; 31; 41. (7.) 38; 46; 54; 62. (8.) 29; 39; 41; 51. (9.) 62; 64; 74; 86. (10.) 63; 77; 89; 91. (11.) 71; 83; 95; 107. (12.) 61; 79; 89; 113.

LESSON XXVII.

(1.) $\frac{1}{2}$. (2.) $\frac{1}{3}$. (3.) $\frac{1}{2}$. (4.) $\frac{2}{3}$. (5.) $\frac{1}{2}$. (6.) $\frac{1}{2}$; $\frac{1}{3}$; $\frac{3}{4}$; $\frac{2}{5}$; $\frac{4}{5}$; $\frac{5}{6}$. (7.) $\frac{1}{2}$; $\frac{1}{3}$; $\frac{2}{3}$ $\frac{3}{4}$; $\frac{4}{5}$; $\frac{5}{6}$. (8.) $\frac{2}{3}$; $\frac{3}{4}$; $\frac{2}{5}$; $\frac{3}{5}$; $\frac{5}{6}$; $\frac{3}{7}$. (9.) $\frac{3}{8}$; $\frac{2}{7}$; $\frac{3}{4}$; $\frac{3}{5}$; $\frac{5}{6}$; $\frac{4}{7}$. (10.) $\frac{4}{9}$; $\frac{2}{3}$; $\frac{3}{4}$; $\frac{3}{5}$; $\frac{5}{6}$; $\frac{4}{7}$. (11.) $\frac{2}{5}$; $\frac{2}{3}$; $\frac{3}{4}$; $\frac{3}{5}$; $\frac{5}{7}$; $\frac{5}{8}$. (12.) $\frac{2}{3}$; $\frac{4}{5}$; $\frac{3}{4}$; $\frac{5}{8}$; $\frac{5}{9}$; $\frac{5}{6}$. (13.) $\frac{3}{4}$; $\frac{4}{5}$; $\frac{5}{6}$; $\frac{6}{7}$; $\frac{7}{9}$; $\frac{7}{8}$. (14.) $\frac{4}{5}$; $\frac{5}{6}$; $\frac{6}{7}$; $\frac{7}{9}$; $\frac{7}{8}$; $\frac{8}{9}$.

LESSON XXVIII.

(1.) $\frac{2}{4}$. (2.) $\frac{2}{6}$. (3.) $\frac{2}{8}$. (4.) $\frac{2}{10}$. (5.) $\frac{2}{12}$. (6.) $\frac{3}{9}$. (7.) $\frac{3}{12}$. (8.) $\frac{4}{6}$. (9.) $\frac{6}{9}$; $\frac{8}{12}$. (10.) $\frac{6}{8}$; $\frac{9}{12}$. (11.) $\frac{4}{10}$; $\frac{6}{15}$. (12.) $\frac{12}{20}$; $\frac{15}{25}$. (13.) $\frac{24}{30}$; $\frac{28}{35}$. (14.) $\frac{10}{12}$; $\frac{15}{18}$. (15.) $\frac{6}{14}$; $\frac{9}{21}$. (16.) $\frac{20}{28}$; $\frac{25}{35}$. (17.) $\frac{6}{16}$; $\frac{9}{24}$. (18.) $\frac{20}{32}$; $\frac{25}{40}$. (19.) $\frac{4}{18}$; $\frac{6}{27}$. (20.) $\frac{16}{36}$; $\frac{20}{45}$. (21.) $\frac{30}{54}$; $\frac{35}{63}$. (22.) $\frac{14}{20}$; $\frac{21}{30}$. (23.) $\frac{10}{24}$; $\frac{15}{36}$.

LESSON XXIX.

(1.) $\frac{8}{12}$ and $\frac{9}{12}$. (2.) $\frac{3}{6}$ and $\frac{2}{6}$; $\frac{5}{10}$ and $\frac{2}{10}$; $\frac{5}{15}$ and $\frac{3}{15}$. (3.) $\frac{4}{12}$ and $\frac{3}{12}$; $\frac{5}{15}$ and $\frac{3}{15}$; $\frac{5}{20}$ and $\frac{4}{20}$. (4.) $\frac{10}{15}$ and $\frac{6}{15}$; $\frac{10}{15}$ and $\frac{9}{15}$; $\frac{10}{15}$ and $\frac{12}{15}$. (5.) $\frac{15}{20}$ and $\frac{8}{20}$; $\frac{15}{20}$ and $\frac{12}{20}$; $\frac{15}{20}$ and $\frac{16}{20}$. (6.) $\frac{12}{30}$ and $\frac{25}{30}$; $\frac{18}{30}$ and $\frac{25}{30}$; $\frac{24}{30}$ and $\frac{25}{30}$. (7.) $\frac{35}{42}$ and $\frac{12}{42}$; $\frac{35}{42}$ and $\frac{18}{42}$; $\frac{35}{42}$ and $\frac{24}{42}$. (8.) $\frac{40}{56}$ and $\frac{21}{56}$; $\frac{48}{56}$ and $\frac{35}{56}$; $\frac{63}{72}$ and $\frac{16}{72}$. (9.) $\frac{40}{90}$ and $\frac{27}{90}$; $\frac{80}{90}$ and $\frac{63}{90}$; $\frac{70}{90}$ and $\frac{81}{90}$. (10.) $\frac{15}{30}$, $\frac{10}{30}$ and $\frac{6}{30}$; $\frac{20}{60}$, $\frac{15}{60}$ and $\frac{12}{60}$. (11.) $\frac{40}{60}$, $\frac{45}{60}$ and $\frac{24}{60}$; $\frac{18}{30}$, $\frac{25}{30}$ and $\frac{15}{30}$. (12.) $\frac{40}{60}$, $\frac{45}{60}$, $\frac{48}{60}$ and $\frac{30}{60}$. (13.) $\frac{9}{12}$ and $\frac{10}{12}$. (14.) $\frac{2}{4}$ and $\frac{1}{4}$; $\frac{3}{6}$ and $\frac{1}{6}$; $\frac{2}{6}$ and $\frac{1}{6}$. (15.) $\frac{4}{6}$ and $\frac{5}{6}$; $\frac{6}{8}$ and $\frac{3}{8}$; $\frac{6}{9}$ and $\frac{2}{9}$. (16.) $\frac{20}{24}$ and $\frac{15}{24}$; $\frac{15}{18}$ and $\frac{14}{18}$; $\frac{21}{24}$ and $\frac{10}{24}$. (17.) $\frac{6}{12}$, $\frac{4}{12}$ and $\frac{3}{12}$; $\frac{8}{24}$, $\frac{6}{24}$ and $\frac{4}{24}$. (18.) $\frac{16}{24}$, $\frac{18}{24}$ and $\frac{20}{24}$; $\frac{16}{24}$, $\frac{18}{24}$ and $\frac{15}{24}$. (19.) $\frac{18}{24}$, $\frac{20}{24}$ and $\frac{21}{24}$, $\frac{12}{18}$, $\frac{15}{18}$ and $\frac{4}{18}$. (20.) $\frac{27}{36}$, $\frac{16}{36}$ and $\frac{15}{36}$; $\frac{30}{36}$, $\frac{32}{36}$ and $\frac{21}{36}$. (21.) $\frac{6}{12}$, $\frac{4}{12}$, $\frac{3}{12}$ and $\frac{2}{12}$; $\frac{10}{30}$, $\frac{6}{30}$, $\frac{5}{30}$ and $\frac{3}{30}$. (22.) $\frac{16}{24}$, $\frac{18}{24}$, $\frac{20}{24}$ and $\frac{21}{24}$; $\frac{12}{36}$, $\frac{9}{36}$, $\frac{20}{36}$ and $\frac{21}{36}$. (23.) $\frac{18}{24}$, $\frac{20}{24}$, $\frac{21}{24}$ and $\frac{16}{24}$; $\frac{72}{180}$, $\frac{120}{180}$, $\frac{120}{180}$ and $\frac{75}{180}$. (24.) $\frac{12}{24}$, $\frac{8}{24}$, $\frac{6}{24}$, $\frac{4}{24}$ and $\frac{3}{24}$; $\frac{16}{24}$, $\frac{18}{24}$, $\frac{20}{24}$, $\frac{21}{24}$ and $\frac{22}{24}$. (25.) $\frac{30}{60}$, $\frac{40}{60}$, $\frac{45}{60}$, $\frac{48}{60}$, $\frac{50}{60}$, $\frac{54}{60}$ and $\frac{55}{60}$.

LESSON XXX.

(1.) $\frac{3}{4}$. (2.) $\$1\frac{1}{4}$. (3.) $\$1\frac{1}{6}$. (4.) $\frac{7}{8}$ orange. (5.) $\frac{7}{12}$. (6.) $\$\frac{3}{5}$. (7.) $3\frac{7}{12}$ yards. (8.) $11\frac{1}{6}$ acres. (9.) $\$1\frac{1}{4}$. (10.) $\frac{4}{5}$; $\frac{7}{8}$; $1\frac{5}{12}$. (11.) $\frac{11}{15}$; $\frac{5}{12}$; $\frac{9}{20}$. (12.) $\frac{9}{14}$; $1\frac{13}{28}$; $1\frac{3}{8}$. (13.) $1\frac{19}{30}$; $1\frac{13}{72}$; $1\frac{17}{63}$. (14.) $4\frac{1}{12}$; $1\frac{17}{60}$. (15.) $8\frac{1}{2}$. (16.) $9\frac{16}{21}$; $11\frac{1}{4}$. (17.) $9\frac{59}{72}$; $13\frac{19}{24}$.

LESSON XXXI.

(1.) $\frac{1}{6}$ orange. (2.) $\frac{1}{4}$ apple. (3.) $\frac{1}{6}$ melon. (4.) $\$\frac{7}{12}$. (5.) $\frac{1}{3}$. (6.) $\frac{7}{12}$ orange. (7.) $\frac{3}{10}$; $\frac{1}{12}$; $\frac{1}{20}$. (8.) $\frac{5}{18}$; $\frac{1}{10}$; $\frac{5}{12}$. (9.) $\frac{4}{15}$; $\frac{1}{40}$; $\frac{5}{21}$. (10.) $\frac{5}{18}$; $\frac{7}{15}$; $\frac{5}{28}$. (11.) $\frac{4}{35}$; $\frac{1}{28}$; $\frac{7}{30}$. (12.) $\frac{1}{24}$; $\frac{1}{42}$; $2\frac{1}{6}$. (13.) $2\frac{1}{2}$. (14.) $4\frac{1}{10}$; $4\frac{7}{12}$. (15.) $2\frac{8}{21}$; $2\frac{1}{18}$; $1\frac{13}{15}$.

LESSON XXXII.

(1.) $\frac{5}{12}$ quart. (2.) $\frac{1}{6}$ apple. (3.) $\frac{1}{10}$ orange. (4.) $\frac{1}{3}$ in oats. (5.) $\frac{4}{9}$; 32 miles. (6.) $\frac{1}{4}$ pound. (7.) $\frac{2}{3}$ bushel. (8.) $\frac{1}{8}$ pound. (9.) $4\frac{1}{12}$ yards. (10.) $\$\frac{3}{16}$. (11.) $\frac{1}{20}$. (12.) 9 sheep. (13.) $\frac{7}{24}$. (14.) $\frac{1}{6}$. (15.) $\frac{1}{6}$. (16.) \$96.

LESSON XXXIII.

(1.) $1\frac{1}{2}$ oranges. (2.) $2\frac{1}{2}$ pecks. (3.) $1\frac{1}{3}$ oranges. (4.) $1\frac{1}{3}$ pineapples. (5.) $2\frac{2}{3}$. (6.) $1\frac{1}{2}$ apples. (7.) $1\frac{1}{2}$ pints. (8.) $2\frac{1}{4}$ oranges. (9.) $4\frac{1}{2}$; $5\frac{1}{4}$; 6. (10.) $\frac{3}{5}$; 2; $3\frac{3}{5}$. (11.) $\frac{3}{7}$; $1\frac{1}{7}$; $2\frac{6}{7}$. (12.) $\frac{5}{8}$; $\frac{3}{4}$; $3\frac{1}{2}$. (13.) $\frac{2}{9}$; $\frac{8}{9}$; $2\frac{2}{9}$. (14.) $2\frac{2}{3}$; 4; $6\frac{2}{9}$. (15.) $5\frac{5}{6}$; $4\frac{4}{5}$; $3\frac{3}{5}$. (16.) $4\frac{3}{8}$; $4\frac{2}{3}$; $3\frac{3}{7}$. (17.) $3\frac{3}{4}$; $4\frac{4}{5}$; $6\frac{6}{7}$. (18.) $11\frac{2}{3}$; 20; 12.

LESSON XXXIV.

(1.) $\$\frac{2}{3}$. (2.) $\$1\frac{1}{3}$. (3.) $\$5\frac{1}{3}$. (4.) $\$4\frac{1}{2}$. (5.) $1\frac{3}{5}$¢. (6.) $4\frac{2}{3}$¢. (7.) $14\frac{2}{5}$¢. (8.) $22\frac{1}{2}$¢. (9.) $6\frac{3}{4}$. (10.) $4\frac{4}{5}$; $5\frac{3}{5}$; $6\frac{2}{5}$; $7\frac{1}{5}$; $8\frac{4}{5}$. (11.) $4\frac{1}{6}$; $5\frac{5}{6}$; $8\frac{1}{3}$; $9\frac{1}{6}$; 10. (12.) $6\frac{6}{7}$; $7\frac{5}{7}$; $8\frac{4}{7}$, $9\frac{3}{7}$, $10\frac{2}{7}$. (13.) $4\frac{3}{8}$; $6\frac{1}{8}$; $7\frac{7}{8}$; $9\frac{5}{8}$; $10\frac{1}{2}$. (14.) $5\frac{1}{3}$; $6\frac{2}{9}$; $7\frac{1}{9}$; $8\frac{8}{9}$; $10\frac{2}{3}$. (15.) $5\frac{2}{5}$; $7\frac{1}{5}$; $8\frac{1}{10}$; $9\frac{9}{10}$, $10\frac{4}{5}$. (16.) $2\frac{8}{11}$; $4\frac{6}{11}$; $7\frac{3}{11}$; $8\frac{2}{11}$; $10\frac{10}{11}$. (17.) $6\frac{5}{12}$; $7\frac{1}{3}$; $8\frac{1}{4}$; $9\frac{1}{6}$; $10\frac{1}{12}$. (18.) $6\frac{2}{5}$ days. (19.) $14\frac{6}{11}$ days. (20.) $4\frac{3}{8}$ days. (21.) $3\frac{3}{7}$ days. (22.) $8\frac{4}{5}$ hours. (23.) $50\frac{7}{8}$¢. (24.) $11\frac{1}{3}$ minutes. (25.) $25\frac{1}{5}$ bushels. (26.) $7\frac{6}{7}$ bushels. (27.) $\$14\frac{2}{5}$. (28.) $5\frac{1}{7}$ feet in the earth; $12\frac{6}{7}$ feet in the air. (29.) A got $\$3\frac{1}{3}$; B got $\$5$; C got $\$6\frac{2}{3}$. (30.) 8¢. (31.) $\$7\frac{1}{5}$. (32.) $\$35$.

LESSON XXXV.

(1.) 3 oranges. (2.) $7\frac{1}{2}$. (3.) $\$2\frac{2}{3}$. (4.) 4; $5\frac{1}{3}$. (5.) 10; 18. (6.) $13\frac{1}{3}$; 22. (7.) $26\frac{2}{3}$; 42. (8.) $\$3\frac{3}{4}$; $\$5$. (9.) $6\frac{1}{4}$; $7\frac{1}{2}$. (10.) $3\frac{1}{2}$; $6\frac{3}{4}$. (11.) 13; $18\frac{3}{4}$. (12.) $19\frac{1}{2}$; 30. (13.) $15\frac{3}{4}$; $24\frac{3}{4}$. (14.) $17\frac{1}{2}$; $32\frac{1}{2}$. (15.) 45. (16.) $9\frac{3}{5}$ barrels. (17.) $13\frac{3}{5}$; 18. (18.) $13\frac{1}{5}$; $8\frac{2}{5}$. (19.) $25\frac{1}{5}$; $22\frac{4}{5}$. (20.) $30\frac{4}{5}$; $27\frac{1}{5}$. (21.) $16\frac{1}{5}$; $28\frac{4}{5}$. (22.) $13\frac{1}{2}$. (23.) 19. (24.) 22. (25.) $25\frac{1}{3}$. (26.) $32\frac{1}{2}$. (27.) 40. (28.) $47\frac{5}{6}$. (29.) $29\frac{5}{7}$. (30.) $37\frac{6}{7}$. (31.) $47\frac{1}{7}$. (32.) $33\frac{1}{2}$. (33.) $42\frac{1}{2}$. (34.) $53\frac{1}{4}$. (35.) $28\frac{2}{3}$. (36.) $48\frac{1}{3}$. (37.) $69\frac{2}{9}$. (38.) $20\frac{3}{5}$. (39.) 53. (40.) $64\frac{4}{5}$. (41.) $96\frac{3}{10}$. (42.) $98\frac{8}{9}$. (43.) $141\frac{9}{11}$. (44.) $57\frac{3}{4}$. (45.) $53\frac{1}{5}$. (46.) 54. (47.) $119\frac{1}{6}$. (48.) $128\frac{8}{9}$.

LESSON XXXVI.

(1.) 4 barrels. (2.) $5\frac{3}{5}$ boxes. (3.) $6\frac{1}{2}$ reams. (4.) $4\frac{2}{3}$. (5.) $6\frac{1}{3}$; $3\frac{4}{5}$; $3\frac{1}{6}$; $2\frac{3}{8}$; $2\frac{1}{9}$; $1\frac{9}{10}$. (6.) $10\frac{1}{4}$; $8\frac{1}{5}$; $5\frac{6}{7}$; $5\frac{1}{8}$; $4\frac{5}{9}$; $4\frac{1}{10}$. (7.) $6\frac{3}{4}$; $4\frac{1}{2}$; $3\frac{3}{8}$; 3; $2\frac{7}{10}$. (8.) 13; $10\frac{5}{6}$; $9\frac{2}{7}$; $7\frac{2}{9}$; $6\frac{1}{2}$. (9.) $8\frac{4}{5}$; $5\frac{1}{2}$; $4\frac{8}{9}$; $4\frac{2}{5}$. (10.) 9; $7\frac{5}{7}$; $6\frac{3}{4}$; 6; $5\frac{2}{5}$. (11.) $11\frac{1}{16}$; $9\frac{4}{7}$; $7\frac{4}{9}$; $6\frac{7}{10}$; $6\frac{1}{11}$. (12.) $10\frac{3}{5}$; $8\frac{5}{6}$; $6\frac{5}{8}$; $5\frac{8}{9}$; $5\frac{3}{10}$. (13.) $4\frac{5}{6}$; $4\frac{1}{7}$; $3\frac{5}{8}$; $3\frac{2}{9}$; $2\frac{9}{10}$ (14.) $11\frac{2}{5}$; $8\frac{1}{7}$; $7\frac{1}{8}$; $5\frac{7}{10}$; $5\frac{2}{11}$. (15.) $10\frac{2}{5}$; $8\frac{2}{3}$; $5\frac{7}{9}$; $5\frac{1}{5}$; $4\frac{1}{3}$. (16.) 2 hundred-weights. (17.) 2 dozen.

(18.) 2 chickens. (19.) 5 dozen. (20.) $10\frac{4}{7}$ pounds. (21.) $5\frac{1}{2}$ barrels. (22.) $6\frac{1}{3}$ pounds. (23.) 10 dozen.

LESSON XXXVII.

(1.) $\frac{1}{4}$ orange. (2.) $\frac{1}{6}$ apple. (3.) $\frac{1}{8}$. (4.) $\frac{1}{9}$. (5.) $\frac{1}{10}$. (6.) $\frac{1}{12}$. (7.) $\frac{1}{14}$. (8.) $\frac{1}{15}$; $\frac{1}{16}$; $\frac{1}{18}$; $\frac{1}{20}$; $\frac{1}{21}$. (9.) $\frac{1}{24}$; $\frac{1}{24}$; $\frac{1}{25}$; $\frac{1}{27}$; $\frac{1}{28}$; $\frac{1}{30}$. (10.) $\frac{1}{32}$; $\frac{1}{35}$; $\frac{1}{36}$; $\frac{1}{49}$; $\frac{1}{64}$; $\frac{1}{81}$. (11.) $\frac{3}{8}$. (12.) $\frac{3}{10}$. (13.) $\frac{2}{15}$. (14.) $\frac{5}{18}$; $\frac{3}{16}$; $\frac{1}{6}$; $\frac{1}{8}$. (15.) $\frac{2}{21}$; $\frac{5}{56}$; $\frac{5}{63}$; $\frac{4}{35}$. (16.) $\frac{1}{18}$; $\frac{5}{81}$; $\frac{7}{90}$; $\frac{8}{99}$. (17.) $\frac{8}{15}$. (18.) $\frac{1}{2}$. (19.) $\frac{2}{5}$; $\frac{9}{10}$; $\frac{5}{9}$; $\frac{3}{14}$. (20.) $\frac{9}{40}$; $\frac{6}{35}$; $\frac{10}{21}$; $\frac{7}{15}$. (21.) $\frac{8}{15}$; $\frac{10}{77}$; $\frac{15}{56}$; $\frac{20}{63}$. (22.) $\frac{5}{4}$ or $1\frac{1}{4}$; $\frac{4}{15}$; $\frac{21}{20}$ or $1\frac{1}{20}$; $\frac{7}{6}$ or $1\frac{1}{6}$. (23.) $\frac{63}{40}$ or $1\frac{23}{40}$; $\frac{88}{21}$ or $4\,\frac{4}{21}$; $\frac{55}{72}$; $\frac{72}{55}$ or $1\frac{17}{55}$. (24.) $\frac{5}{8}$. (25.) $\frac{12}{35}$. (26.) $\frac{7}{45}$. (27.) $\frac{16}{33}$. (28.) $\frac{10}{21}$.

LESSON XXXVIII.

(1.) $1\frac{1}{4}$ bushels. (2.) $\frac{5}{6}$; $\frac{5}{8}$; $\frac{7}{12}$; $\frac{11}{20}$; $\frac{13}{20}$. (3.) $\frac{7}{6}$ or $1\frac{1}{6}$; $\frac{31}{42}$; 1; $\frac{5}{4}$ or $1\frac{1}{4}$; $\frac{8}{21}$. (4.) $\frac{26}{15}$ or $1\frac{11}{15}$; $\frac{48}{35}$ or $1\frac{13}{35}$; $\frac{35}{16}$ or $2\frac{3}{16}$; $\frac{51}{64}$; $\frac{41}{21}$ or $1\frac{20}{21}$. (5.) $\$2$. (6.) $\$3\,\frac{2}{3}$ (7.) $\$2\frac{3}{4}$. (8.) $\$\frac{32}{49}$. (9.) $\$2\frac{3}{16}$. (10.) $\$2\frac{2}{5}$; $\$3\frac{1}{5}$. (11.) $\frac{5}{2}$. (12.) $\frac{7}{5}$. (13.) $\$6\frac{2}{9}$; $\$7\frac{7}{9}$. (14.) $5\frac{5}{21}$ days. (15.) 9 days.

LESSON XXXIX.

(1.) $\$40$. (2.) 6 marbles. (3.) $\$25$. (4.) 16 marbles. (5.) $\$35$. (6.) $\$32$. (7.) $\$35$. (8.) $\$21$. (9.) $\$27$. (10.) $\$50$; $\$30$. (11.) 35¢. (12.) 42 trees. (13.) 30 feet. (14.) 30 feet. (15.) $17\frac{1}{2}$ feet. (16.) 35 feet. (17.) $\$6$. (18.) $\$10$. (19.) 10 barrels. (20.) $6\frac{3}{4}$ barrels. (21.) $2\frac{1}{2}$ yards. (22.) $2\frac{4}{7}$ yards. (23.) $\$1\frac{4}{5}$. (24.) 6 lemons. (25.) 6 oranges. (26.) 8 barrels. (27.) 2 barrels. (28.) 20 yards.

LESSON XL.

(1.) $4\frac{1}{5}$. (2.) $5\frac{1}{3}$. (3.) $6\frac{2}{9}$. (4.) $7\frac{7}{8}$. (5.) $11\frac{4}{7}$. (6.) $3\frac{3}{5}$. (7.) $6\frac{3}{7}$. (8.) 12. (9.) $8\frac{3}{4}$. (10.) $10\frac{4}{5}$. (11.) 3. (12.) $3\frac{1}{3}$ bushels. (13.) $3\frac{1}{3}$ apples. (14.) 2. (15.) $5\frac{1}{4}$. (16.) 8. (17.) $6\frac{3}{10}$. (18.) $5\frac{1}{7}$. (19.) $6\frac{7}{8}$. (20.) $3\frac{1}{3}$. (21.) $2\frac{2}{3}$. (22.) $8\frac{1}{10}$. (23.) $6\frac{2}{3}$. (24.) $4\frac{16}{21}$. (25.) $9\frac{5}{8}$. (26.) $6\frac{3}{7}$. (27.) $8\frac{3}{4}$. (28.) $9\frac{3}{5}$. (29.) $6\frac{3}{4}$. (30.) $5\frac{5}{6}$. (31.) $5\frac{2}{5}$. (32.) $5\frac{9}{11}$.

LESSON XLI.

(1.) $\frac{2}{3}$ yards. (2.) $\frac{3}{4}$ yards. (3.) $\frac{2}{5}$ yards. (4.) $\frac{3}{5}$ bushel. (5.) $1\frac{1}{8}$ yards.

(6.) $1\frac{2}{3}$ pounds. (7.) $\frac{5}{6}$ pounds. (8.) $1\frac{7}{8}$ yards; $2\frac{1}{12}$ yards; $2\frac{1}{2}$ yards. (9.) $1\frac{1}{15}$ bushels; $\frac{4}{3}$ bushels. (10.) $\frac{4}{5}$. (11.) $\frac{5}{6}$; $\frac{2}{3}$; $\frac{1}{2}$. (12.) $\frac{6}{7}$; $\frac{5}{7}$; $\frac{4}{7}$. (13.) $\frac{7}{8}$; $\frac{3}{4}$; $1\frac{1}{8}$. (14.) $1\frac{1}{5}$. (15.) $1\frac{1}{15}$. (16.) $1\frac{1}{24}$; $1\frac{1}{9}$; $1\frac{1}{4}$. (17.) $1\frac{1}{35}$; $1\frac{1}{14}$; $1\frac{1}{7}$. (18.) $1\frac{1}{48}$; $1\frac{1}{20}$; $1\frac{3}{32}$. (19.) $1\frac{1}{6}$. (20.) $1\frac{1}{63}$; $1\frac{1}{27}$; $1\frac{1}{15}$. (21.) $1\frac{1}{80}$; $1\frac{1}{35}$; $1\frac{1}{20}$. (22.) $1\frac{1}{99}$; $1\frac{1}{44}$; $1\frac{3}{77}$. (23.) $1\frac{1}{120}$; $1\frac{1}{54}$; $1\frac{1}{32}$. (24.) $1\frac{4}{5}$; $1\frac{5}{9}$; $1\frac{3}{7}$. (25.) $2\frac{2}{15}$; $1\frac{31}{32}$; $2\frac{1}{4}$. (26.) $1\frac{2}{9}$; $1\frac{8}{25}$; $1\frac{19}{77}$. (27.) $1\frac{2}{3}$; 2; $1\frac{11}{25}$. (28.) $1\frac{1}{11}$; $2\frac{7}{9}$; $4\frac{2}{5}$.

LESSON XLII.

(1.) $1\frac{5}{7}$ acres. (2.) $\frac{2}{5}$. (3.) $\frac{3}{5}$. (4.) $\frac{3}{4}$. (5.) $\frac{7}{4}$ or $1\frac{3}{4}$; $\frac{3}{4}$. (6.) $\frac{3}{7}$; $\frac{5}{7}$. (7.) 4¢. (8.) $\$3$. (9.) 7 miles. (10.) $\$15$. (11.) $\$20$; $\$5$. (12.) $\$4\frac{1}{2}$. (13.) $\$3\frac{3}{4}$. (14.) $\$1\frac{7}{8}$. (15.) $2\frac{2}{5}$. (16.) $4\frac{2}{3}$. (17.) $2\frac{1}{4}$.

LESSON XLIII.

(1.) $7\frac{2}{7}$. (2.) 3 bushels. (3.) 4 persons. (4.) 15 yards. (5.) $4\frac{1}{3}$ yards. (6.) 5 apples; 7 apples. (7.) $6\frac{1}{2}$ yards. (8.) $\frac{1}{2}$ times; $\frac{8}{15}$ times; $1\frac{5}{6}$ times. (9.) $\frac{10}{27}$ times; $\frac{20}{63}$ times; $1\frac{13}{18}$ times. (10.) $\frac{15}{128}$ times; $\frac{15}{112}$ times; $1\frac{3}{4}$ times. (11.) $10\frac{2}{3}$; $26\frac{2}{3}$. (12.) $23\frac{1}{4}$; $54\frac{1}{4}$. (13.) $15\frac{7}{15}$; $11\frac{3}{5}$. (14.) $11\frac{2}{3}$; $5\frac{3}{5}$. (15.) $4\frac{8}{9}$; $6\frac{1}{9}$. (16.) 23 times; 31 times; 28 times. (17.) 4 times; $7\frac{2}{3}$ times; $10\frac{1}{3}$ times. (18.) $7\frac{14}{15}$ times; $10\frac{8}{9}$ times; $18\frac{1}{12}$ times. (19.) $7\frac{3}{5}$ times; $8\frac{24}{25}$ times; $13\frac{5}{7}$ times. (20.) $3\frac{9}{20}$ times; $10\frac{5}{22}$ times; $14\frac{1}{8}$ times; 16 times. (21.) 6 gallons; $10\frac{1}{2}$ gallons. (22.) 6 bushels; 11 bushels.

LESSON XLIV.

(1.) $\$18$. (2.) $\$2$. (3.) $\$3\frac{1}{3}$. (4.) $\$1.04$. (5.) $\$4$; $\$\frac{1}{5}$. (6.) $\$2\frac{7}{10}$. (7.) $\$11\frac{7}{10}$; $\$\frac{3}{20}$. (8.) $\$25$. (9.) $\$9\frac{3}{8}$; $\$\frac{3}{16}$. (10.) $\$\frac{1}{4}$. (11.) $\$\frac{1}{15}$. (12.) $\$\frac{1}{10}$; $\$3$. (13.) $\$7\frac{1}{2}$. (14.) $\$2$. (15.) $\$2$. (16.) $\$9$. (17.) $\$10$. (18.) $\$5$. (19.) $\$18$. (20.) $\$25$. (21.) 36 pounds. (22.) 10 dozen. (23.) 12 yards. (24.) 12 yards. (25.) 400. (26.) 900; 2500; 5100. (27.) 3600; 900; 1700. (28.) 2600; 5600; 6000. (29.) 12. (30.) 12; 28; 10. (31.) $9\frac{1}{3}$; 12; 12.

LESSON XLV.

(1.) 34¢. (2.) $\$43$. (3.) 8¢. (4.) $\$8$. (5.) $\$64$. (6.) 35¢. (7.) $\$19$. (8.) $\$26$. (9.) 9¢. (10.) $\$24$. (11.) $\$13$. (12.) $\$78$. (13.) $\$42$; $\$30$.

(14.) \$16; \$36. (15.) \$49. (16.) \$56. (17.) 12¢. (18.) 9 apples. (19.) 54¢. (20.) \$49. (21.) 48¢; 64¢; 80¢. (22.) 21 miles. (23.) 18 rows. (24.) \$63. (25.) 20 bushels. (26.) 7 weeks. (27.) 10 apples. (28.) 8 horses.

LESSON XLVI.

(1.) \$20; \$36. (2.) 64; 81. (3.) 8 oranges. (4.) 16 times. (5.) 8 chestnuts. (6.) 6 3/4 times. (7.) 77; 82. (8.) 2/5; 2/7; 5/6. (9.) 48/144; 36/144; 34/144. (10.) 50 1/8 acres. (11.) 3 17/24. (12.) 5/12; 35 miles. (13.) 3 1/2. (14.) 7 7/11. (15.) \$26 1/4. (16.) 54. (17.) 3 33/49. (18.) 14 quarts. (19.) 2 1/2. (20.) \$1 19/20. (21.) \$72. (22.) 10 barrels. (23.) 8 times. (24.) 9. (25.) 64. (26.) 4. (27.) \$4. (28.) 24 apples. (29.) 10 11/15. (30.) \$6 3/10. (31.) \$3 3/4. (32.) 23 1/4 yards. (33.) \$1 1/2. (34.) \$2/3. (35.) 20 lemons. (36.) 2 yards. (37.) father—50 years; son—5 years. (38.) 45 bushels. (39.) \$3 1/3. (40.) \$8 4/5. (41.) 24. (42.) 12. (43.) 2. (44.) 18. (45.) 40. (46.) 10. (47.) 12¢. (48.) 6¢. (49.) 1 pear. (50.) \$4. (51.) 10 peaches. (52.) 10 chestnuts. (53.) \$6. (54.) 3 1/3 quarts. (55.) 3¢. (56.) \$12; \$4. (57.) 1/4. (58.) 1/9. (59.) 5/16. (60.) 1/3. (61.) \$20. (62.) 32 yards. (63.) 12 days. (64.) \$48. (65.) 14 days. (66.) 20 days. (67.) 7 1/2 days. (68.) 4 apples. (69.) 26. (70.) 5. (71.) 5. (72.) 10 1/6. (73.) 6. (74.) 4¢. (75.) 15 marbles to Thomas; 6 marbles. (76.) \$5. (77.) \$54. (78.) \$28. (79.) \$1. (80.) 8 days. (81.) 30 seconds; hound—300 ft.; hare—210 ft. (82.) 20 rods. (83.) 5 hours; 25 miles. (84.) 8 hours. (85.) 12 hours. (86.) \$32. (87.) 10¢. (88.) \$3 3/4. (89.) 2/3. (90.) 1 1/11 days. (91.) \$15; \$3/7. (92.) 8 bushels. (93.) 24 hours.

LESSON XLVII.

(1.) 20 mills. (2.) 30 mills; 40 mills; 50 mills; 60 mills; 70 mills; 80 mills; 90 mills. (3.) 20 cents; 30 cents; 40 cents; 50 cents; 60 cents; 70 cents; 80 cents; 90 cents. (4.) 20 dimes; 30 dimes; 40 dimes; 50 dimes; 60 dimes; 70 dimes; 80 dimes; 90 dimes. (5.) \$20; \$30; \$40; \$50; \$60; \$70; \$80; \$90. (6.) 200 ¢. (7.) 300¢; 400¢; 500¢; 600¢; 700¢; 800¢; 900¢. (8.) 2 dimes. (9.) 3 dimes; 4 dimes; 5 dimes; 6 dimes; 7 dimes; 8 dimes; 9 dimes. (10.) \$2; \$3; \$4; \$5; \$6; \$7; \$8; \$9. (11.) \$2. (12.) \$3; \$4; \$5; \$6; \$7; \$8; \$9.

LESSON XLVIII.

(1.) 4 pints; 6 pints; 8 pints; 10 pints; 12 pints; 14 pints. (2.) 16 quarts; 24 quarts. (3.) 8 pecks; 12 pecks; 16 pecks; 20 pecks; 24 pecks; 28 pecks; 32 pecks; 36 pecks; 40 pecks; 44 pecks; 48 pecks. (4.) 2 quarts; 3 quarts; 4 quarts; 5 quarts; 6 quarts; 7 quarts. (5.) 2 pecks; 3 pecks. (6.) 2 bushels; 3 bushels; 4 bushels; 5 bushels; 6 bushels; 7 bushels. (7.) 7 pints. (8.) 29 quarts. (9.) 14 pecks. (10.) 39 pints. (11.) 95 quarts. (12.) 101 pints. (13.) 3 quarts 1 pint. (14.) 4 qt. 1 pt.; 5 qt. 1 pt.; 6 qt. 1 pt.; 7 qt. 1 pt. (15.) 1 pk. 2 qt.; 1 pk. 3 qt.; 1 pk. 4 qt.; 1 pk. 5 qt.; 2 pk. 1 qt.; 2 pk. 3 qt.; 2 pk. 7 qt.; 3 pk. 3 qt. (16.) 1 bu. 3 pk.; 2 bu. 2 pk.; 3 bu. 1 pk.; 3 bu. 3 pk.; 4 bu. 1 pk.; 5 bu. 3 pk. (17.) 1 pk. 5 qt. 1 pt. (18.) 2 pk. 1 qt. 1 pt.; 2 pk. 3 qt. 1 pt.; 2 pk. 5 qt. 1 pt.; 2 pk. 6 qt. 1 pt. (19.) 1 bu. 2 pk. 5 qt.; 1 bu. 2 pk. 7 qt.; 1 bu. 3 pk. 1 qt.; 1 bu. 3 pk. 3 qt. (20.) 2 bu. 2 pk. 3 qt.; 2 bu. 2 pk. 6 qt.; 2 bu. 3 pk. 1 qt.; 2 bu. 3 pk. 4 qt. (21.) 3 bu. 1 pk. 7 qt. 1 pt.; 3 bu. 2 pk.; 3 bu. 2 pk. 1 qt.; 3 bu. 2 pk. 2 qt. (22.) 5 bu. 1 pk. 4 qt. 1 pt.; 5 bu. 1 pk. 5 qt.; 5 bu. 1 pk. 5 qt. 1 pt.; 5 bu. 1 pk. 6 qt. (23.) 194 quarts. (24.) 129 pints. (25.) 288 pints. (26.) 207 pints. (27.) 369 pints. (28.) 510 pints.

LESSON XLIX.

(1.) 4 pints; 6 pints. (2.) 8 qt.; 12 qt.; 16 qt.; 20 qt.; 24 qt.; 28 qt.; 32 qt.; 36 qt.; 40 qt.; 44 qt.; 48 qt. (3.) 1 pt. 1 gi.; 1 pt. 2 gi.; 1 pt. 3 gi. (4.) 1 qt. 1 pt.; 2 qt.; 2 qt, 1 pt.; 3 qt.; 3 qt. 1 pt. (5.) 1 gal. 1 qt.; 1 gal. 2 qt.; 1 gal. 3 qt.; 2 gal.; 2 gal. 1 qt.; 2 gal. 2 qt.; 2 gal. 3 qt.; 3 gal. (6.) 8 gi.; 16 gi.; 24 gi. (7.) 8 pt.; 16 pt.; 24 pt.; 32 pt.; 40 pt. (8.) 32 gi.; 64 gi.; 96 gi.; 128 gi.; 160 gi. (9.) 7 gi. (10.) 7 pt. (11.) 22 qt. (12.) 21 gi. (13.) 39 pt. (14.) 47 gi. (15.) 1 qt. 1 pt. 1 gi. (16.) 2 gal. 3 qt. 1 pt. (17.) 2 gal. 1 qt. 1 pt. 3 gi. (18.) 3 gal. 1 gi.; 3 gal. 2 gi.; 3 gal. 1 pt.; 3 gal. 1 pt. 2 gi. (19.) 5 gal. 3 qt. 3 gi.; 5 gal. 3 qt. 1 pt.; 5 gal. 3 qt. 1 pt. 2 gi.; 5 gal. 4 qt. (20.) 200 gi. (21.) 259 gi. (22.) 440 gi.

LESSON L.

(1.) 32 dr.; 48 dr.; 64 dr.; 80 dr.; 96 dr. (2.) 32 oz.; 48 oz.; 64 oz.;

80 oz.; 96 oz.; 160 oz. (3.) 200 lb.; 300 lb.; 400 lb.; 500 lb.; 600 lb.; 700 lb.; 800 lb.; 900 lb. (4.) 40 cwt.; 60 cwt.; 80 cwt.; 100 cwt.; 120 cwt. (5.) 1 oz. 4 dr.; 1 oz. 11 dr.; 2 oz. 3 dr. (6.) 2 lb. 10 oz.; 3 lb. 5 oz.; 4 lb. 11 oz.; 5 lb. 10 oz. (7.) 3 cwt.; 4 cwt. 50 lb.; 5 cwt. 75 lb.; 12 cwt. (8.) 2 T.; 2 T. 10 cwt.; 3 T. 15 cwt.; 4 T.; 4 T. 16 cwt. (9.) 8 oz. (10.) 4 oz.; 12 oz. (11.) 50 lb.; 25 lb.; 75 lb.; 20 lb.; 40 lb.; 60 lb.; 80 lb. (12.) 75 dr. (13.) 2731 dr. (14.) 1545 lb. (15.) 9075 lb. (16.) $\frac{3}{4}$ lb. (17.) $\frac{1}{2}$ lb.; $\frac{5}{8}$ lb.; $\frac{7}{8}$ lb. (18.) $\frac{1}{10}$; $\frac{1}{5}$; $\frac{1}{4}$; $\frac{2}{5}$; $\frac{1}{2}$; $\frac{3}{5}$; $\frac{3}{4}$; $\frac{4}{5}$. (19.) $\frac{1}{5}$; $\frac{1}{4}$; $\frac{3}{10}$; $\frac{2}{5}$; $\frac{1}{2}$; $\frac{3}{5}$; $\frac{3}{4}$; $\frac{4}{5}$; $\frac{9}{10}$.

LESSON LI.

(1.) 24 in. (2.) 6 ft.; 9 ft.; 12 ft.; 15 ft. (3.) 11 yd.; 22 yd.; 27$\frac{1}{2}$ yd.; 38 $\frac{1}{2}$ yd.; 55 yd. (4.) 120 rods; 160 rods; 200 rods; 240 rods; 280 rods; 360 rods. (5.) 72 fur. (6.) 3 ft.; 4 ft. (7.) 5 yd.; 7 yd. (8.) 4 rods; 6 rods. (9.) 2 mi. (10.) 2 ft. 6 in. (11.) 8 in.; 9 in. (12.) 2 ft.; 2 ft. 3 in. (13.) 3 yd.; 2 yd. 2 ft. 3 in.; 3 yd. 2 ft.; 2 yd. 7$\frac{1}{5}$ in. (14.) 2 fur.; 2 fur. 26 rd. 3 yd. 2 ft.; 5 fur. 13 rd. 1 yd. 2 ft. 6 in.; 6 fur.; 3 fur. 8 rd. (15.) $\frac{1}{20}$ mi. (16.) $\frac{3}{40}$ mi. (17.) $\frac{6}{11}$ rd. (18.) $\frac{2}{3}$ yd.; $\frac{4}{33}$ rd. (19.) $\frac{1}{2}$ ft.; $\frac{1}{6}$ yd.; $\frac{1}{33}$ rd.

LESSON LII.

(1.) 30 sec.; 20 sec.; 40 sec.; 45 sec. (2.) 12 min.; 24 min.; 36 min .; 48 min.; 10 min.; 50 min.; 8 min. 34$\frac{2}{7}$ sec. (3.) 12 hr.; 16 hr.; 20 hr.; 6 hr. 51 min. 25$\frac{5}{7}$ sec.; 9 hr.; 16 hr. 48 min. (4.) 3 da.; 5 da. (5.) 2 wk.; 3 wk.; 1 wk. 1 da.; 3 wk. 3 day.; 2 wk. 4 da; 3 wk. 5 da. (6.) $\frac{1}{3}$ min. (7.) $\frac{5}{6}$ hr. (8.) $\frac{1}{2}$ da.; $\frac{1}{14}$ wk. (9.) $\frac{5}{7}$ wk. (10.) $\frac{2}{3}$ yr. (11.) $\frac{7}{48}$ da. (12.) $\frac{1}{2}$ wk. (13.) $\frac{5}{14}$ mon. (14.) $\frac{8}{21}$ mon. (15.) 61 da. (16.) 92 da. (17.) 91 da. (18.) 16 da. (19.) 59 da. (20.) 186 da. (21.) 183 da.

LESSON LIII.

(1.) \$3.45. (2.) \$1.85. (3.) 77$\frac{1}{2}$¢. (4.) \$66. (5.) 85 mi. (6.) 2 gal. 3 qt. 1 pt. (2$\frac{7}{8}$ gal.) (7.) 16 lb. 10$\frac{2}{3}$ oz. (16$\frac{2}{3}$ lb.) (8.) 7$\frac{3}{5}$ hr. (9.) 7$\frac{3}{4}$ ft. (2 yd. 1$\frac{3}{4}$ ft.) (10.) $\frac{3}{4}$ pk. (11.) 95¢. (12.) 23 yd. 1 ft. (4 rd. 1 yd. 1 ft.) (13.) \$2$\frac{1}{4}$ (\$2.25) (14.) 25 mph. (15.) 3 rev. (16.) 13 wk. (17.) 5 lb. 14 oz. (18.) \$18. (19.) 60¢. (20.) \$22.50; 15¢. (21.) \$90$\frac{10}{11}$.

LESSON LIV.

(1.) 165 gal. (2.) \$50. (3.) \$185. (4.) \$19. (5.) \$35. (6.) 3¢ (7.) \$2. (8.) \$12.50. (9.) 140 miles. (10.)\$2.82. (11.) 66¢. (12.) 600 yd. (13.) 28 mi. (14.) 40 lb. (15.) \$11. (16.) $22\frac{1}{5}$ lb. (17.) 100 bu.

LESSON LV.

(1.) $\$1\frac{1}{2}$ (\$1.50) (2.) $\$5\frac{5}{8}$. (3.) \$5. (4.) 15¢. (5.) $\$13\frac{1}{2}$ (\$13.50) (6.) 35 trees. (7.) \$25. (8.) $\$8\frac{3}{4}$. (9.) 51¢. (10.) 58 mi. (11.) $\$1\frac{3}{8}$. (12.) $\$12\frac{3}{5}$. (13.) $\$5\frac{1}{3}$. (14.) $9\frac{1}{3}$. (15.) $3\frac{3}{4}$. (16.) $\$3\frac{3}{5}$. (17.) $\$\frac{3}{4}$. (18.) $2\frac{4}{5}$. (19.) $3\frac{3}{8}$. (20.) $5\frac{1}{2}$. (21.) $5\frac{5}{9}$. (22.) $1\frac{3}{5}$. (23.) $\frac{7}{12}$. (24.) $2\frac{2}{15}$. (25.) 11 marbles. (26.) 24 plums. (27.) 8 mi. (28.) \$25. (29.) \$30. (30.) 14 plums. (31.) 20¢. (32.) A had \$28; B had \$16. (33.) 32 apples; 24 pears; 8 plums; 32 cherries. (34.) 35 apples; 10 pears.

LESSON LVI.

(1.) 7. (2.) 8. (3.) 4 and 12. (4.) 6 and 42. (5.) 4, 8 and 12. (6.) 5, 15 and 25. (7.) 30. (8.) 5. (9.) 3 and 9. (10.) 12. (11.) 4. (12.) 13. (13.) 6 and 10. (14.) 10 and 15. (15.) 12 and 19. (16.) A has 6¢, B has 9¢. (17.) Thomas has \$12, James has \$7. (18.) 12¢. (19.) Thomas had 11, William had 9. (20.) Mary had 17, Sarah had 7. (21.) 15. (22.) 28. (23.) 35. (24.) Henry is 10; Oliver is 13; James is 14. (25.) Jane has 10¢; Sarah has 15¢; Mary has 18¢. (26.) Frank is 15; Mary is 27. (27.) chain—\$15; ring—\$20; watch—\$27. (28.) 126¢. (29.) James had \$7; John had \$17; Frank had \$31. (30.) Thomas has \$4; Joseph has \$10; Paul has \$8. (31.) Harness cost \$25; horse cost \$75; buggy cost \$125.

LESSON LVII.

(1.) 6 and 9. (2.) Thomas owes \$42; John owes \$18. (3.) First day—\$32; second day—\$24. (4.) 37 and 63. (5.) 20, 10, 15. (6.) A has 20¢; B has \$12; C has \$8. (7.) 30, 25, 15. (8.) Hat cost \$8; coat cost \$20; vest cost \$6. (9.) A's share was 18¢; B's share was 20¢. (10.) 20 cows; 35 sheep. (11.) 24 and 36. (12.) Sarah is 6; Mary is 8. (13.) 24 and 27. (14.) 30 apple trees; 35 peach trees. (15.) 6 miles. (16.) 36 apple trees; 12 plum trees; 21 cherry trees. (17.) Sarah is 24; Jane is 21. (18.) 9. (19.) Charles has \$60; John has \$36.

(20.) 48 hogs; 32 sheep; 24 cows. (21.) 4 o'clock. (22.) 4:30. (23.) 2 o'clock. (24.) 3 o'clock. (25.) 4 o'clock. (26.) 1:20.

LESSON LVIII.

(1.) 10. (2.) 12. (3.) 15. (4.) 20. (5.) 28. (6.) 20 (7.) 12. (8.) 24. (9.) 50¢. (10.) 30. (11.) The son is 15; the father is 55. (12.) 15. (13.) 36 yd. (14.) A to B is 15 mi.; C to D is 30 mi. (15.) 30.

LESSON LIX.

(1.) $\frac{1}{2}$. (2.) $\frac{1}{10}$. (3.) 2 times. (4.) $2\frac{2}{3}$ lots. (5.) $\frac{2}{5}$. (6.) $\frac{2}{7}$. (7.) $\frac{3}{5}$. (8.) $\frac{3}{4}$. (9.) $\frac{19}{20}$. (10.) $1\frac{1}{5}$ da. (11.) 4 da. (12.) $2\frac{11}{12}$ da. (13.) 1 da. (14.) 6 da. (15.) 20 da. (16.) 24 da. (17.) $1\frac{1}{5}$ da. (18.) $1\frac{3}{7}$. (19.) 2 da.

LESSON LX.

(1.) $\frac{1}{4}$; $\frac{1}{2}$; $\frac{1}{8}$. (2.) 5; $\frac{1}{5}$. (3.) 6 to 1. (4.) 2; 2 to 1. (5.) 3 to 1; 5 to 1; 6 to 1; 4 to 1; 1 to 2; 3 to 4. (6.) 1 to 2; 1 to 2; 1 to 2; 4 to 5; 5 to 6; 3 to 2. (7.) 30. (8.) 12. (9.) 2. (10.) 2. (11.) $2\frac{1}{2}$. (12.) John gets 10¢; George gets 15¢. (13.) 20. (14.) A gets 8; B gets 12. (15.) John get 12¢; James get 16¢. (16.) 60 apple trees; 36 peach trees. (17.) 14 boys; 21 girls. (18.) 12. (19.) 5 and 20. (20.) 12 and 16. (21.) A pays $20; B pays $25. (22.) First gets 5 doz.; second gets $2\frac{1}{2}$ doz. (23.) A gets $35; B gets $21. (24.) C lost $18; D lost $12.

LESSON LXI.

(1.) 10 and 12. (2.) Henry gets 6 apples; Oliver gets 10 apples. (3.) A gets 6¢; B gets 8¢. (4.) James has 21 marbles; John has 12 marbles. (5.) First got $10; second got $18. (6.) William is 20; Frank is 12. (7.) 21 sound apples; 9 apples which are not sound. (8.) One built 12 feet; the other built 15 feet. (9.) 15, 20 and 25. (10.) 7, 14, 21 and 28. (11.) 18, 12 and 9. (12.) William got 9 peaches; Thomas got 12 peaches; John got 15 peaches. (13.) A lost $45.50; B lost $91; C lost $227.50. (14.) A has $24; B has $12; Charles has $6. (15.) A gets 20¢; B gets 15¢; C gets 10¢. (16.) 6 horses; 18 cows; 36 sheep. (17.) A gets 7 plums; B gets 14 plums; C gets 21 plums. (18.) Emma gets 5 cherries; Agnes gets 10 cherries, Sarah gets 20 cherries.

LESSON LXII.

(1.) 10 men. (2.) 10 men. (3.) 8 da. (4.) $4\frac{1}{2}$ hr. (5.) $48. (6.) $12. (7.) $48. (8.) 4. (9.) 10. (10.) 15; 12. (11.) 6 oz.; $4\frac{4}{5}$ oz. (12.) 12 oz. (13.) 8 oz. (14.) 50 men. (15.) $4\frac{1}{6}$ da. (16.) $5\frac{1}{2}$ da. (17.) $56. (18.) $45. (19.) $7\frac{1}{2}$ rd. (20.) 27 bu. (21.) 10 da. (22.) $880. (23.) 126 sheep.

LESSON LXIII.

(1.) A should pay $15, and B $10. (2.) A should pay $35, and B $25. (3.) A should pay $24, and B $48. (4.) A should pay $14, and B $21. (5.) C should get $18, and D $27. (6.) A should get $36, and B $45. (7.) A should get $25, and B $30. (8.) E should pay $9, and F $18. (9.) M should get $60, and N $90. (10.) C should get $150, and D $100. (11.) E's share $210 and F's $560. (12.) A invested $1300, and B $1100. (13.) C invested $420 and D $560. (14.) A invested $700, and B $1000. (15.) E gained $300, and F $540.

LESSON LXIV.

(1.) $\frac{1}{50}$. (2.) $\frac{1}{25}$; $\frac{1}{20}$; $\frac{3}{50}$; $\frac{2}{25}$. (3.) $\frac{1}{10}$; $\frac{3}{25}$; $\frac{3}{20}$. (4.) $\frac{4}{25}$; $\frac{1}{5}$; $\frac{6}{25}$. (5.) $\frac{1}{4}$; $\frac{7}{25}$; $\frac{3}{10}$. (6.) $\frac{8}{25}$; $\frac{7}{20}$; $\frac{9}{25}$. (7.) $\frac{2}{5}$; $\frac{9}{20}$; $\frac{12}{25}$. (8.) $\frac{1}{2}$; $\frac{3}{5}$; $\frac{7}{10}$. (9.) $\frac{3}{4}$; $\frac{4}{5}$; $\frac{9}{10}$. (10.) $\frac{1}{40}$. (11.) $\frac{7}{200}$; $\frac{1}{16}$. (12.) $\frac{1}{15}$; $\frac{3}{40}$. (13.) $\frac{1}{12}$; $\frac{1}{8}$. (14.) $\frac{2}{15}$; $\frac{1}{6}$. (15.) $\frac{7}{40}$; $\frac{3}{16}$. (16.) $\frac{7}{30}$; $\frac{5}{16}$. (17.) $\frac{3}{8}$; $\frac{7}{16}$. (18.) $\frac{9}{16}$; $\frac{5}{8}$. (19.) $\frac{2}{3}$. (20.) $\frac{7}{8}$.

LESSON LXV.

(1.) 2. (2.) 3; 6. (3.) 2; 3. (4.) 3; 6. (5.) $8; $20. (6.) 17 bu. (7.) 7 horses. (8.) $2. (9.) $1. (10.) 38 sheep. (11.) 55 sheep. (12.) $3.80. (13.) 33¢. (14.) 9¢; 18¢. (15.) $6\frac{1}{2}$¢; $19\frac{1}{2}$¢. (16.) 15¢; 20¢; 25¢; $43\frac{3}{4}$¢.

LESSON LXVI.

(1.) 50%. (2.) $33\frac{1}{3}$%; $66\frac{2}{3}$%; 25%. (3.) 75%; 20%; 40%. (4.) 60%; 80%; $16\frac{2}{3}$%. (5.) $12\frac{1}{2}$%; $37\frac{1}{2}$%; $62\frac{1}{2}$%. (6.) 10%; 30%; 70%. (7.) 90%; $8\frac{1}{3}$%; $41\frac{2}{3}$%. (8.) $6\frac{2}{3}$%; $6\frac{1}{4}$%; $18\frac{3}{4}$%. (9.) $31\frac{1}{4}$%; 5%; 15%. (10.) 35%; 45%; 55%. (11.) 4%; 8%; 12%. (12.) 16%; 24%; 28%. (13.) 32%; 36%; 44%. (14.) $23\frac{1}{3}$%; $5\frac{5}{7}$%; $22\frac{1}{2}$%. (15.) $27\frac{1}{2}$%; $13\frac{1}{3}$%; $10\frac{5}{12}$%. (16.) 2%; 6%; 14%. (17.) 18%; $1\frac{2}{3}$%; $1\frac{1}{3}$%.

(18.) 62$\frac{1}{2}$%; 75%; 87$\frac{1}{2}$%. (19.) 56$\frac{1}{4}$%; 68$\frac{3}{4}$%; 81$\frac{1}{4}$%. (20.) 16$\frac{2}{3}$%; 50%; 83$\frac{1}{3}$%. (21.) 41$\frac{2}{3}$%; 58$\frac{1}{3}$%; 68$\frac{3}{4}$%. (22.) 91$\frac{2}{3}$%; 93$\frac{3}{4}$%; 15%.

LESSON LXVII.

(1.) 40%. (2.) 60%; 25%. (3.) 50%; 12$\frac{1}{2}$%. (4.) 25%; 16$\frac{2}{3}$%. (5.) 5%. (6.) 25%. (7.) 33$\frac{1}{3}$%. (8.) 66$\frac{2}{3}$%. (9.) 20%. (10.) 40% (11.) 25%. (12.) 20%. (13.) 50%. (14.) 60%. (15.) 30%. (16.) 60%. (17.) 60%. (18.) 16$\frac{2}{3}$%; 25%. (19.) 55$\frac{5}{9}$%.

LESSON LXVIII.

(1.) \$10. (2.) \$20. (3.) 10¢. (4.) \$72. (5.) \$70. (6.) \$12. (7.) 3¢. (8.) \$8; \$4 (or 50%). (9.) 25%. (10.) 50%. (11.) \$1.50. (12.) \$2. (13.) \$4. (14.) 60%. (15.) 16$\frac{2}{3}$%. (16.) 20%. (17.) 140. (18.) 12. (19.) 3$\frac{11}{20}$. (20.) 8.

LESSON LXIX.

(1.) \$2$\frac{2}{5}$. (2.) \$100. (3.) \$700. (4.) \$380. (5.) \$950. (6.) \$6. (7.) \$1080. (8.) \$175. (9.) \$480. (10.) \$55.10.

LESSON LXX.

(1.) \$100. (2.) \$28. (3.) \$12.50. (4.) \$75; \$1425. (5.) \$980. (6.) \$2000. (7.) \$600. (8.) 19 shares.

LESSON LXXI.

(1.) \$10. (2.) \$30. (3.) \$17.50. (4.) \$30. (5.) \$17. (6.) \$37.50.

LESSON LXXII.

(1.) 30¢. (2.) 60¢. (3.) \$2. (4.) \$4.80. (5.) \$6. (6.) \$8. (7.) \$9. (8.) \$8.40. (9.) \$9. (10.) \$36.

LESSON LXXIII.

(1.) \$1.50. (2.) \$1. (3.) \$2.80. (4.) \$2.40. (5.) \$4.50. (6.) \$3.90. (7.) \$6.20. (8.) \$3.53. (9.) \$4.80. (10.) \$19.20. (11.) \$8.45. (12.) \$4.24. (13.) \$3.20. (14.) \$6.75. (15.) \$3.80. (16.) \$28. (17.) \$44. (18.) \$68.20. (19.) \$32.80. (20.) \$56.80. (21.) \$99.12.

LESSON LXXIV.

(1.) $25. (2.) $50. (3.) $60. (4.) $75. (5.) $140. (6.) $240. (7.) $350. (8.) $4000.

LESSON LXXV.

(1.) $50. (2.) $200. (3.) $500. (4.) $250. (5.) $300. (6.) $25. (7.) $125.

LESSON LXXVI.

(1.) 3 yr. 4 mo. (2.) 4 yr. (3.) 2 yr. 6 mo. (4.) 2 yr. 8 mo. (5.) 3 yr. 5 mo. 4$\frac{2}{7}$ da. (6.) 6 yr. 8 mo. (7.) 25 yr. (8.) 50 yr.; 33$\frac{1}{3}$ yr.; 20 yr.; 16$\frac{2}{3}$ yr.; 14$\frac{2}{7}$ yr.; 12$\frac{1}{2}$ yr.; 10 yr.; 8$\frac{1}{3}$ yr. (9.) 40 yr. (10.) 25 yr.; 20 yr.

LESSON LXXVII.

(1.) 6%. (2.) 8%. (3.) 5%. (4.) 7%. (5.) 8%. (6.) 7%. (7.) 5%. (8.) 6%. (9.) 5%. (10.) 8$\frac{1}{3}$%; 10%; 12$\frac{1}{2}$%; 20%; 25%; 50%.

LESSON LXXVIII.

(1.) Present worth is $60; discount is $12. (2.) Present worth is $400; discount is $120. (3.) Present worth is $25; discount is $5. (4.) Present worth is $500; discount is $250. (5.) The discount is $45. (6.) The discount is $96. (7.) The discount is $4. (8.) The present worth is $50. (9.) The present worth is $44. (10.) The present worth is $55. (11.) The discount is $196.79. (12.) The discount is $162.30.

LESSON LXXIX.

(1.) $\frac{1}{4}$. (2.) $\frac{1}{5}$. (3.) 10%. (4.) 10%. (5.) 6 yr. (6.) 25%. (7.) 40%. (8.) 6%. (9.) $16. (10.) $15.30. (11.) 8 yr. (12.) A has $160; B has $80. (13.) A has $300; B has $200. (14.) A has $177.78 (or 177\frac{7}{9}$; B has $133.33 (or 133\frac{1}{3}$).

LESSON LXXX.

(1.) 105 plums. (2.) Mary gets 13 peaches; James gets 8 peaches; Lucy gets 11 peaches. (3.) 8. (4.) A has $3; B has $6; C has $18.

(5.) James has $16; Thomas has $18. (6.) 72 sheep. (7.) 2½ da. (8.) 4¢. (9.) A should pay $16; B should pay $36; C should pay $40. (10.) 8 children. (11.) 10. (12.) 3 children. (13.) 60 leaps. (14.) 140 steps. (15.) $60. (16.) A had 14¢; B had 10¢. (17.) A is 5; B is 10; C is 30. (18.) A has $90; B has $75. (19.) 10 hrs. (20.) 25%. (21.) 7 playmates. (22.) 105 steps. (23.) The watch cost $35; the chain $10. (24.) 6 da. (25.) 80. (26.) 9%. (27.) 10 yd. (28.) A has $40; B has $11. (29.) $25. (30.) 60 leaps. (31.) Thomas is 15; James is 5. (32.) John traveled 56 mi.; George traveled 30 mi. (33.) I would make an additional 11¢ per yard (instead of making 4¢ per yard I would make 15¢ per yard). (34.) 7⅗¢ per dozen. (35.) A is 20; B is 10. (36.) 73 mi. (37.) A can do the job in 12 da.; B in 24 da.; C in 8 da. (38.) 30 ducks; 60 chickens. (39.) 50¢. (40.) 11¼ mi. (41.) $25. (42.) A received $480; B received $560; C received $640. (43.) The man would consume the meal in 24 da; his wife in 40 da. (44.) 20 lb. (45.) A is 50; B is 10; C is 5. (46.) Mary is 9; Ella is 15. (47.) A can do the job in 24 days; B in 48 days. (48.) 12¢. (49.) 5 da. (50.) 22 da. (51.) 16 yd. (52.) 100 steps. (53.) 14 men (assuming a full 7 day week). (54.) First field contains 10 sheep; second field contains 40 sheep; third field contains 120 sheep. (55.) 20 da. (56.) A get $12; B gets $18; C get $24. (57.) 48¢. (58.) $200. (59.) 17 lb. (60.) 9 steps. (61.) John should get 6¢; James 3¢.

SOLUTIONS

OF THE

MORE DIFFICULT EXAMPLES

IN

RAY'S NEW INTELLECTUAL ARITHMETIC.

LESSON I.

18. All together have the sum of 5 dollars, 3 dollars, and 1 dollar, which is 9 dollars.

25. As many as the sum of 4, 4, and 2, which is 10.

LESSON II.

25. I have the sum of 10 cents, 5 cents, 3 cents, and 3 cents, which is 21 cents.

LESSON V.

37. William's age is the sum of 8 years and 5 years, which is 13 years; and the sum of all their ages is 13 years, plus 5 years, plus 8 years, which is 26 years.

LESSON VI.

24. I received in money as many dollars as the difference between 17 dollars and 6 dollars, which is 11 dollars.

LESSON VIII.

19. I spent the sum of 20 cents and 10 cents, which is 30 cents; I had left the difference between 65 cents and 30 cents, which is 35 cents.

LESSON IX.

8. Both had the sum of 18 marbles and 18 marbles, which is 36 marbles; if when they quit one had 25 marbles, the other had the difference between 36 marbles and 25 marbles, which is 11 marbles.

14. He is worth the sum of 20 dollars and 10 dollars, which is 30 dollars. He owes the sum of 5 dollars, 6 dollars, and 10 dollars, which is 21 dollars. Should he pay his debts, he would be worth the difference between 30 dollars and 21 dollars, which is 9 dollars.

LESSON XI.

4. The sum of 19 and 10 is 29; the difference between 17 and 10 is 7; if I take 7, the difference, from 29, the sum, 22 will be left.

21. There are as many peach-trees in the orchard as the sum of 15 peach-trees, 9 peach-trees, and 10 peach-trees, which is 34 peach-trees. There are as many apple-trees as the sum of 5 apple-trees, 11 apple-trees, and 10 apple-trees, which is 26 apple-trees. Then there are as many more peach-trees than apple-trees as the difference between 34 trees and 26 trees, which is 8 trees.

LESSON XII.

22. If 1 yard of muslin cost 11 cents, 3 yards will cost 3 times 11 cents, which is 33 cents.

LESSON XIV.

11. If a man travel 7 miles in 1 hour, in 8 hours he will travel 8 times 7 miles, which is 56 miles.

17. In one hour they would be as far apart as the sum of 2 miles and 4 miles, which is 6 miles; in 3 hours they would be 3 times 6 miles apart, which is 18 miles.

LESSON XVI.

22. Each one would receive one sixth of 36 dollars, which is 6 dollars.

23. Since there are 4 quarts in 1 gallon, in 36 quarts there are as many gallons as 4 quarts are contained times in 36 quarts, which are 9.

LESSON XVIII.

24. One man will earn in 3 days one ninth of $108, which is $12. In one day he would earn one third of $12, which is $4..

25. In 1 day the former travels one third of 15 miles, which is 5 miles. In 1 day the latter travels one half of 20 miles, which is 10 miles; and if the latter travels 10 miles in 1 day, and the former, 5 miles, the latter travels as much farther in 1 day than the former as the difference between 10 miles and 5 miles, which is 5 miles.

LESSON XIX.

7. The sum of 1, 2, and 3 is 6, and 6 is contained in 60 ten times. If I have as many marbles as 3 times the number of times 6 is contained in 60, I have 3 times 10 marbles, which is 30 marbles.

8. Six hats will cost 6 times $5, which is $30; 4 yards of cloth will cost 4 times $3, which is $12. Both will cost the sum of $30 and $12, which is $42; and if he gave in exchange flour at $6 a barrel, it took as many barrels as $6 are contained times in $42, which are 7.

9. If a man gain 6 miles in 5 hours, it will take as many times 5 hours to gain 24 miles as 6 miles are contained times in 24 miles, which are 4; and 4 times 5 hours are 20 hours.

34. It will take 1 man 3 times 10 days, which is 30 days. It will take as many men to do it in 5 days as 5 days are contained times in 30 days, which are 6.

LESSON XXII.

2. One third of an apple is worth $\frac{1}{3}$ of three cents, which is 1 cent.

5. One fourth of a melon is worth $\frac{1}{4}$ of 8 cents, which is 2 cents; and if $\frac{1}{4}$ of a melon is worth 2 cents, $\frac{3}{4}$ of a melon are worth 3 times 2 cents, which is 6 cents.

23. One calf cost $\frac{1}{12}$ of \$120, which is \$10; and if 1 calf cost \$10, he sold the 7 calves for 7 times \$10, which is \$70.

LESSON XXIII.

9. For \$1 you can buy $\frac{1}{8}$ of a bushel, and for \$5 you can buy 5 times $\frac{1}{8}$ of a bushel, which is $\frac{5}{8}$ of a bushel.

16. One fifth of 30 is 6; then $\frac{3}{5}$ of 30 are 3 times 6, which is 18; and 18 is $\frac{18}{23}$ of 23.

LESSON XXIV.

7. One sixth of a gallon will cost $\frac{1}{5}$ of 35 cents, which is 7 cents; and if $\frac{1}{6}$ of a gallon cost 7 cents, $\frac{6}{6}$, or 1 gallon, will cost 6 times 7 cents, which is 42 cents.

17. One fourth of 8 cents is 2 cents, and $\frac{3}{4}$ are 3 times 2 cents, which is 6 cents; and if 6 cents are $\frac{2}{3}$ of mine, $\frac{1}{3}$ of mine is $\frac{1}{2}$ of 6 cents, which is 3 cents; and if 3 cents are $\frac{1}{3}$, then $\frac{3}{3}$ will be 3 times 3 cents, which is 9 cents.

LESSON XXV.

4. One yard will cost $\frac{1}{3}$ of 5 dollars, which is $1\frac{2}{3}$ dollars.

LESSON XXVI.

4. Since there are $\frac{4}{4}$ in 1, in 3 there are 3 times $\frac{4}{4}$, which is $\frac{12}{4}$; and $\frac{12}{4}+\frac{1}{4}=\frac{13}{4}$. Other answers, $\frac{19}{4}$, $\frac{21}{4}$, $\frac{27}{4}$.

LESSON XXVII.

13. To reduce a fraction to its lowest terms, divide both terms by their greatest common divisor. Of 27 and 36, the G. C. D. is 9; 9 in 27 is contained 3 times, and 9 in 36 is contained 4 times.

Therefore, $\frac{27}{36}$ changed to its lowest terms $=\frac{3}{4}$.

LESSON XXVIII.

10. Since there are $\frac{8}{8}$ in 1, in $\frac{1}{4}$ there is $\frac{1}{4}$ of $\frac{8}{8}$, which is $\frac{2}{8}$; and if $\frac{2}{8}=\frac{1}{4}$, then $\frac{3}{4}$ will be 3 times $\frac{2}{8}$, which is $\frac{6}{8}$.

LESSON XXIX.

4. The common denominator is 15. $1=\frac{15}{15}$; $\frac{1}{3}=\frac{5}{15}$, and $\frac{2}{3}=\frac{10}{15}$; $\frac{1}{5}=\frac{3}{15}$, and $\frac{2}{5}=\frac{6}{15}$.

LESSON XXX.

2. Three fourths $=\frac{6}{8}$, and $\frac{1}{2}=\frac{4}{8}$. He gave for both $\$\frac{6}{8}+\$\frac{4}{8}=\$\frac{10}{8}=\$1\frac{1}{4}$.

LESSON XXXI.

3. One half of the first $=\frac{3}{6}$ of a melon; $\frac{2}{3}$ of the second $=\frac{4}{6}$ of a melon. $\frac{4}{6}-\frac{3}{6}=\frac{1}{6}$.

LESSON XXXII.

14. Find how much is in both air and water: as much as the sum of $\frac{1}{2}$ and $\frac{1}{3}$. $\frac{1}{2}=\frac{3}{6}$, and $\frac{1}{3}=\frac{2}{6}$; their sum is $\frac{5}{6}$. Since there are $\frac{6}{6}$ in the pole, there would be as much in the earth as the difference between $\frac{6}{6}$ and $\frac{5}{6}$, which is $\frac{1}{6}$.

LESSON XXXIII.

2. To 5 horses he would give 5 times $\frac{1}{2}$ peck, which is $\frac{5}{2}$ pecks; and $\frac{5}{2}$ pecks $= 2\frac{1}{2}$ pecks.

LESSON XXXIV.

8. One pound of cheese will sell for $\frac{1}{4}$ of 30 cents, which is $7\frac{1}{2}$ cents. Then 3 pounds will sell for 3 times $7\frac{1}{2}$ cents, which is $22\frac{1}{2}$ cents.

28. One seventh of 18 is $2\frac{4}{7}$; then $\frac{2}{7}=2$ times $2\frac{4}{7}$ feet, which is $5\frac{1}{7}$ feet; and $\frac{5}{7}=5$ times $2\frac{4}{7}$ feet, which is $12\frac{6}{7}$ feet.

29. One ninth of \$15 is \$$1\frac{2}{3}$; then $\frac{2}{9}=2$ times \$$1\frac{2}{3}$, which is \$$3\frac{1}{3}$; $\frac{1}{3}$ of \$15 is \$5; \$5 + \$$3\frac{1}{3}$ = \$$8\frac{1}{3}$, \$15 — \$$8\frac{1}{3}$ = \$$6\frac{2}{3}$.

31. $\frac{5}{5}-\frac{2}{5}=\frac{3}{5}$. \$18 $=\frac{3}{5}$ of the number, then $\frac{1}{5}$ would be $\frac{1}{3}$ of \$18, which is \$6; and $\frac{2}{5}$ would be 2 times \$6, which is \$12.

32. \$45 $=\frac{7}{7}+\frac{2}{7}$, which is $\frac{9}{7}$ of the cost. $\frac{1}{7}$ is $\frac{1}{9}$ of \$45 = \$5; and $\frac{7}{7}$ would be 7 times \$5 = \$35.

LESSON XXXVI.

22. Five and three sevenths pounds of sugar cost $5\frac{3}{7}$ times 7 cents, which is 38 cents. It would take as many pounds of raisins to pay for it as 6 cents are contained times in 38 cents, which are $6\frac{1}{3}$.

LESSON XXXVIII.

9. One pound will cost $\frac{1}{4}$ of $\$\frac{5}{4} = \$\frac{5}{16}$; then 7 pounds will cost 7 times $\$\frac{5}{16} = \$\frac{35}{16} = \$2\frac{3}{16}$.

LESSON XXXIX.

16. The sum of $\frac{1}{5}$ and $\frac{2}{5} = \frac{3}{5}$; $\frac{5}{5} - \frac{3}{5} = \frac{2}{5}$; then 14 ft. $= \frac{2}{5}$ of the pole; $\frac{1}{5} = 7$ ft.; $\frac{5}{5} = 35$ ft.

23. $12 = $\frac{4}{3}$ of the cost; $\frac{1}{3}$ of the cost is $\frac{1}{4}$ of $12 = $3; $\frac{3}{3} = \$9$. One yard cost $\frac{1}{5}$ of $9 = $$1\frac{4}{5}$.

28. One eighth of the cost was $\frac{1}{5}$ of $50, which is $10; then $\frac{8}{8}$ are 8 times $10, which is $80. It would take as many yards as $4 are contained times in $80, which are 20.

LESSON XL.

11. One third of the number is $\frac{1}{8}$ of 56, which is 7; then $\frac{3}{3}$ are 3 times 7, which is 21; 21 is 3 times 7.

LESSON XLI.

9. One fourth of a bu. of wheat is worth $\frac{1}{3}$ of a bu. of rye; then $\frac{4}{4}$ of a bu. of wheat are worth $\frac{4}{3}$ of a bu. of rye; and $\frac{1}{5}$ of a bu. of wheat is worth $\frac{1}{5}$ of $\frac{4}{3}$ of a bu. of rye, which is $\frac{4}{15}$ of a bu. of rye; and $\frac{4}{5}$ of a bu. of wheat are worth 4 times $\frac{4}{15}$ of a bu. of rye, which is $\frac{16}{15}$, or $1\frac{1}{15}$ bu. of rye.

LESSON XLIII

22. As many bu. of rye as $\frac{3}{4}$ are contained times in $4\frac{1}{2}$. $4\frac{1}{2}=\frac{18}{4}$; $\frac{18}{4}\div\frac{3}{4}=6$.

LESSON XLIV.

11. $\$3\frac{1}{10}=\$\frac{31}{10}$; $7\frac{3}{4}=\frac{31}{4}$; $\frac{1}{4}$ of a doz. cost $\frac{1}{31}$ of $\$\frac{31}{10}$, which is $\$\frac{1}{10}$; then 1 pair cost $\frac{1}{3}$ of $\$\frac{1}{10}$, or $\$\frac{1}{30}$. He gained the difference between $\$\frac{1}{10}$ and $\$\frac{1}{30}$, which is $\$\frac{1}{15}$.

12. $2\frac{1}{2}=\frac{5}{2}$. $\frac{1}{2}$ doz. cost $\frac{1}{5}$ of \$15, which is \$3; then each one cost $\frac{1}{6}$ of \$3, which is $\$\frac{1}{2}$. He gained on each one the difference between $\$\frac{3}{5}$ and $\$\frac{1}{2}$, which is $\$\frac{1}{10}$.

On $\frac{1}{2}$ doz. he gained $\$\frac{6}{10}$; and on $\frac{5}{2}$ doz. $\$\frac{30}{10}$, or \$3.

LESSON XLV.

22. A walks 5 miles 7 times in walking 35 miles; B walks 3 miles 7 times in the same time. Therefore, B walks 7 times 3 miles, which is 21 miles.

25. One horse will eat $\frac{1}{6}$ of 12 bu. in a week, which is 2 bu. a week; then 10 horses will eat 10 times 2 bu. in a week, which is 20 bushels.

26. Five horses will eat in 1 week $\frac{1}{2}$ of 16 bu., which is 8 bu; to eat 56 bu., it will take them as many weeks as 8 is contained times in 56, which are 7.

28. It will take 6 times 12 horses, which is 72 horses, to eat it in 1 day; and to eat it in 9 days it will take $\frac{1}{9}$ of 72 horses, which is 8 horses.

LESSON XLVI.

6. Nine times $9 = 81$. $81 \div 12 = 6\frac{9}{12}$, or $6\frac{3}{4}$.

8. $\frac{48}{120} = \frac{2}{5}$. $\frac{54}{189} = \frac{2}{7}$. $\frac{240}{288} = \frac{5}{6}$.

9. One ninth $= \frac{16}{144}$, $\frac{3}{9} = \frac{48}{144}$; $\frac{1}{16} = \frac{9}{144}$, $\frac{4}{16} = \frac{36}{144}$; $\frac{1}{72} = \frac{2}{144}$, $\frac{17}{72} = \frac{34}{144}$.

12. If he traveled $\frac{1}{4}$, or $\frac{3}{12}$, the first day, and $\frac{1}{3}$, or $\frac{4}{12}$, the second day, then the third day he must have traveled $\frac{12}{12}$ less $\frac{7}{12}$, which is $\frac{5}{12}$; $\frac{5}{12}$ of 84 miles $= 35$ miles.

21. $\$99 = \frac{8}{8} + \frac{3}{8}$, or $\frac{11}{8}$, of the cost; then $\frac{1}{8} = \$9$, and $\frac{8}{8}$, the cost, $= \$72$.

22. One eighth of $\$96 = \$12$, or $\frac{1}{5}$ of the cost; then the cost was 5 times $\$12 = \$60$. It took as many barrels of flour to pay for the horse as \$6 are contained times in \$60, which are 10.

23. Eighty-four is $\frac{7}{6}$ of 72, and 72 is 8 times 9.

25. Eight ninths of $81 = 72$, and $72 = \frac{9}{8}$ of 64.

26. Four sevenths of 35 are 20, and 20 is $\frac{5}{6}$ of 24. Three eighths of 16 are 6, and 24 is 4 times 6.

27. $\$17\frac{1}{2} = \$\frac{35}{2}$. $4\frac{3}{8}$ yd. $= \frac{35}{8}$ yd. $\frac{1}{8}$ of a yd. would cost $\frac{1}{35}$ of $\$\frac{35}{2}$, which is $\$\frac{1}{2}$; then $\frac{8}{8}$ of a yd. would cost $\$\frac{8}{2}$, or \$4.

33. In one week he would earn $\frac{1}{8}$ of \$72, which is \$9; in one day he would earn $\frac{1}{6}$ of \$9, which is $\$1\frac{1}{2}$.

37. One half of 20 years $=$ 10 years, or $\frac{1}{5}$ of the father's age; then 5 times 10 years $=$ 50 years, the father's age. $\frac{1}{10}$ of 50 years $=$ 5 years, or the age of the youngest son.

38. \$21 $= \frac{7}{5}$ of the cost. $\frac{1}{5} = \frac{1}{7}$ of \$21, which is \$3. $\frac{5}{5} = \$15$, or the cost. At \$1 a bushel, it would take 15 bushels of corn to pay for it; at $\$\frac{1}{3}$, it would take 3 times 15 bushels, which is 45 bushels.

39. Three yards $= \frac{15}{5}$ yards. $\frac{15}{5}$ are 5 times $\frac{3}{5}$, and will cost 5 times $\$\frac{2}{3}$, which is $\$3\frac{1}{3}$.

41. One half of 12 $=$ 6. $6 + 2 = 8$. 8 is $\frac{1}{3}$ of 24.

44. Three fourths of 24 $=$ 18. $18 - 6 = 12$. 12 is $\frac{2}{3}$ of 18. 18 is 6 more than $\frac{2}{3}$ of itself.

50. Two fifths of 30 yards $=$ 12 yards. He sold one yard for $\frac{1}{12}$ of \$48, which is \$4.

60. Three fifths of \$20 are \$12. Fourteen is $\frac{7}{9}$ of 18, and 2 times 18 are 36. Twelve is $\frac{1}{3}$ of 36.

79. Two fifths of 10 yards are 4 yards, and they cost $\frac{2}{5}$ of \$90, which is \$36. \$40 \$36 $=$ \$4, the gain on 4 yards, and on 1 yard the gain is \$1.

80. B gains in one day 23 miles less 18 miles, which are 5 miles. It will take as many days to gain 40 miles as 5 miles are contained times in 40 miles, which are 8.

81. The hound gains in one second 10 feet less 7 feet = 3 ft., or 1 yd.; then to gain 90 yards it will take 90 seconds, or $1\frac{1}{2}$ min. The hound runs 90 times 10 feet, which are 900 ft. = 300 yd. The hare runs 90 times 7 ft. = 630 feet, or 210 yd.

85. In one hour the cistern would lose 9 gallons less 6 gallons, which are 3 gallons; and it would take as many hours to empty the cistern as 3 gallons are contained times in 36 gallons, which are 12.

89. Such part of the journey as $2\frac{1}{4}$ days are of $3\frac{3}{8}$ days. $3\frac{3}{8} = \frac{27}{8}$. $2\frac{1}{4} = \frac{18}{8}$. $\frac{18}{8} = \frac{18}{27}$, or $\frac{2}{3}$, of $\frac{27}{8}$. He can therefore perform $\frac{2}{3}$ of the journey in $2\frac{1}{4}$ days.

90. In one day A can do $\frac{1}{2}$, B $\frac{1}{4}$, and C $\frac{1}{6}$; then all do the sum of $\frac{1}{2}$, $\frac{1}{4}$, and $\frac{1}{6}$, which is $\frac{11}{12}$, in one day. To do $\frac{12}{12}$, it will take as many days as $\frac{11}{12}$ are contained times in $\frac{12}{12}$, which are $1\frac{1}{11}$.

91. Twenty yards at \$4 per yard = \$80. 15 yards at \$3 per yard = \$45. \$80 + \$45 = \$125, or what I paid for 35 yards. I received for $\frac{6}{7}$, or 30 yards, \$3 per yard = \$90, and for $\frac{1}{7}$, or 5 yards, \$4 per yard = \$20. For all I received \$90 + \$20 = \$110. My loss on 35 yards was \$125 — \$110 = \$15, or $\$\frac{15}{35} = \$\frac{3}{7}$ per yard.

93. Three fourths of 6 miles are $4\frac{1}{2}$ miles. 6 miles less $4\frac{1}{2}$ miles = $1\frac{1}{2}$ miles, the distance B gains in one hour. To gain 36 miles, it will take as many hours as $1\frac{1}{2}$ miles are contained times in 36 miles, which are 24.

LESSON LIII.

1. Five bu. will cost 5 times 60 cents, which is \$3; 3 pk. will cost $\frac{3}{4}$ of 60 cents, which is 45 cents; then 5 bu. and 3 pk. will cost \$3.45.

2. Four gal. 2 qt. 1 pt. = 37 pt.; at 5 cents a pint, the milk will cost 37 times 5 cents, which is \$1.85.

4. One rod contains 198 in.; 2 yd. 2 ft. 3 in. = 99 in., or $\frac{1}{2}$ a rod; then $5\frac{1}{2}$ rods will cost $5\frac{1}{2}$ times \$12, which is \$66.

5. Twenty-six min. and 40 sec. = $\frac{4}{9}$ of an hour; 9 hours + $\frac{4}{9}$ hours = $9\frac{4}{9}$ hours. If it traveled 9 miles an hour, the distance is $9\frac{4}{9}$ times 9 miles = 85 miles.

8. Three tenths da. = $\frac{3}{10}$ of 24 hr., which is $7\frac{1}{5}$ hr.; $7\frac{1}{5}$ hr. + $\frac{2}{5}$ hr. = $7\frac{3}{5}$ hours.

9. One third rd. = $5\frac{1}{2}$ ft., or $5\frac{2}{4}$ ft.; $\frac{1}{2}$ yd. = $1\frac{1}{2}$ ft., or $1\frac{2}{4}$; then $5\frac{2}{4}$ ft. + $1\frac{2}{4}$ + $\frac{3}{4}$ ft. = $7\frac{3}{4}$ ft.

11. One bu. 3 pk. = 7 pk; 1 pk. is worth $\frac{1}{7}$ of 70 cents, which is 10 cents; 2 bu. 1 pk. = 9 pk.; and 4 qt. = $\frac{1}{2}$ pk.; then $9\frac{1}{2}$ pk. are worth $9\frac{1}{2}$ times 10 cents, which is 95 cents.

13. One third of a T. cost $\frac{1}{2}$ of \$8, which is \$4; 1 T. cost \$12; 1 cwt. cost $\frac{1}{20}$ of \$12, which is 60 ct.; 3 cwt. 75 lb. = $3\frac{3}{4}$ cwt.; then $3\frac{3}{4}$ cwt. cost $3\frac{3}{4}$ times 60 ct., which is \$2.25.

14. In 2 hr. 24 min. there are 144 min.; the rate per min. is $\frac{1}{144}$ of 60 miles = $\frac{60}{144}$ mi., or $\frac{5}{12}$ mi.; the rate per hr. is 60 times $\frac{5}{12}$ mi., which is $\frac{300}{12}$ mi. = 25 mi.

15. In 3 yd. 1 ft. 6 in. there are $3\frac{1}{2}$ yd.; in 1 rd. 5 yd. there are $10\frac{1}{2}$ yd.; the wheel would make as many revolutions in going $10\frac{1}{2}$ yd. as $3\frac{1}{2}$ yd. are contained times in $10\frac{1}{2}$ yd., which are 3.

17. I bought as many pounds as 40 ct. are contained times in 235 ct., which are $5\frac{7}{8}$; $\frac{7}{8}$ lb. $=$ 14 oz. I bought 5 lb. 14 oz.

20. In 150 bu. are 6 T.; 6 T. will cost 6 times \$3.75, which is \$22.50; 1 bu. will cost $\frac{1}{25}$ of \$3.75, which is 15 ct.

21. The distance around the lot is 50 ft. $+$ 100 ft. $\times$ 2 $=$ 300 ft.; 300 ft. $=$ 100 yd.; 100 yd. $= 18\frac{2}{11}$ rd. If 1 rd. cost \$5, then $18\frac{2}{11}$ rd. cost $18\frac{2}{11}$ times \$5, which is $\$90\frac{10}{11}$.

LESSON LIV.

15. The entire cost was \$90 plus $\$3 \times 6 = \$108$; the sum received for him was $\$42 + \$99 = \$141$; all gained $\$141 - \$108 = \$33$; each man received $\frac{1}{3}$ of \$33, which is \$11.

LESSON LV.

12. In 6 da. of 8 hr. each there are 48 hr.; in 7 da. of 9 hr. each there are 63 hr.; $\$9\frac{3}{5} = \$\frac{48}{5}$. In 1 hr. he would earn $\frac{1}{48}$ of $\$\frac{48}{5}$, which is $\$\frac{1}{5}$; in 63 hr., $\$\frac{63}{5} = \$12\frac{3}{5}$.

15. In $3\frac{1}{2}$ are $\frac{7}{2}$; $2\frac{1}{3} = \frac{7}{3}$. $\frac{1}{3}$ of the number is $\frac{1}{7}$ of $\frac{7}{2} = \frac{1}{2}$; then $\frac{3}{3} = \frac{3}{2}$, or $1\frac{1}{2}$. $1\frac{1}{2} \times 2\frac{1}{2} = \frac{3}{2} \times \frac{5}{2} = \frac{15}{4} = 3\frac{3}{4}$.

18. Two thirds of $\frac{6}{5} = \frac{4}{5}$. If $\frac{4}{5}$ are $\frac{2}{7}$, then $\frac{1}{7}$ is $\frac{1}{2}$ of $\frac{4}{5}$, which is $\frac{2}{5}$; and $\frac{7}{7}$ would be 7 times $\frac{2}{5}$, which is $\frac{14}{5} = 2\frac{4}{5}$.

22. Two thirds of $\frac{12}{5}$ are $\frac{8}{5}$. If $\frac{8}{5}$ are $\frac{1}{2}$, then $\frac{2}{2}$ are $\frac{16}{5}$; and 2 is contained in $\frac{16}{5}$, $\frac{8}{5}$ or $1\frac{3}{5}$ times.

25. Four fifths of 10 marbles are 8 marbles. If 8 is $\frac{8}{11}$, then $\frac{1}{11}$ is 1, and $\frac{11}{11}$ are 11.

26. Three fifths of 60 plums are 36 plums; $\frac{3}{4}$ of 36 are 27; $\frac{4}{9}$ of 27 are 12, or what she gave away. She had left $36 - 12 = 24$.

27. Five sevenths of the distance is 35 mi.; then $\frac{1}{7}$ is 7 mi., and $\frac{2}{7}$ are 14 mi.; $\frac{3}{7}$ of 14 mi. = 6 mi.; 14 mi. — 6 mi. = 8 mi.

30. Seven sevenths less $\frac{2}{7} = \frac{5}{7}$; $\frac{2}{5}$ of $\frac{5}{7} = \frac{2}{7}$; $\frac{5}{7} - \frac{2}{7} = \frac{3}{7}$, the part she had left. $\frac{3}{7} = 6$, $\frac{1}{7} = 2$, $\frac{7}{7} = 14$.

31. Two thirds of 12 ct. are 8 ct. If 8 is $\frac{1}{2}$, then $\frac{2}{2}$ are 16; if 16 ct. are $\frac{4}{5}$ of William's money, then William has 20 ct.

32. If $\frac{1}{2}$ of B's money equals $\frac{2}{7}$ of A's, then all of B's money $= \frac{4}{7}$ of A's; $\frac{7}{7} - \frac{4}{7} = \frac{3}{7}$, the difference between A's and B's money; $\frac{3}{7} = 12$ ct., $\frac{1}{7} = 4$ ct., $\frac{7}{7} = 28$ ct., A's money; 28 ct. — 12 ct. = 16 ct., B's.

33. One third $= \frac{4}{12}$, $\frac{1}{4} = \frac{3}{12}$; $\frac{4}{12} + \frac{3}{12} + \frac{1}{12} = \frac{8}{12} = \frac{2}{3}$; then $32 = \frac{1}{3}$; $\frac{3}{3} = 96$, the number of trees in the orchard. $\frac{1}{3}$ of $96 = 32$; $\frac{1}{4}$ of $96 = 24$; $\frac{1}{12}$ of $96 = 8$.

34. If $\frac{2}{9}$ are pear-trees, $\frac{7}{9}$ must be apple-trees. The excess of apple-trees is therefore $\frac{5}{9}$ of the whole; 25 is then $\frac{5}{9}$ of the whole; $\frac{2}{9}$, or the pear-trees, = 10, and $\frac{7}{9}$, or the apple-trees, = 35.

LESSON LVI.

3. If the second is three times the first, then the whole number is four times the first. Therefore, the first is $\frac{1}{4}$ of $16 = 4$, and the second $4 \times 3 = 12$.

5. The whole number will be six times the first part; then the first part $= \frac{1}{6}$, the second $\frac{2}{6}$, the third $\frac{3}{6}$, or 4, 8, and 12, respectively.

10. The difference of the two numbers is $6 + 2 = 8$; the sum of 8, the difference, and 4, one of the numbers, $= 12$, the other number.

11. The sum of 19 and 6 is 25; $25 - 10 = 15$, the difference between the numbers; then $19 - 15 = 4$, the smaller number.

12. The sum of the numbers is $10 + 8 = 18$; $18 - 5 = 13$, the other number.

18. They had at first 32 ct. — 8 ct. = 24 ct.; each had $\frac{1}{2}$ of 24 ct. = 12 ct. If Thomas found 8 more, he had 12 ct. + 8 ct. = 20 ct.

19. They bought 4 peaches + 6 peaches + 20 peaches = 30 peaches; each one bought $\frac{1}{2}$ of 30 peaches, which is 15 peaches. Thomas had left $15 - 4 = 11$; William had left $15 - 6 = 9$.

20. Both bought 24 cherries + 7 cherries + 5 cherries = 36 cherries. Since Mary bought twice as many as Sarah, both bought three times as many as Sarah; there-

fore Sarah bought $\frac{1}{3}$ of 36 cherries $=$ 12 cherries, and Mary bought 2 times 12 cherries $=$ 24 cherries; $24 - 7 = 17$, the number of cherries Mary had left; $12 - 5 = 7$, the number Sarah had left.

21. Three times the number is $50 - 5 = 45$; $\frac{1}{3}$ of 45 is 15, the number.

22. Three fourths of the number would be $31 - 10 = 21$; $\frac{1}{4}$ would be $\frac{1}{3}$ of $21 = 7$; $\frac{4}{4}$, or the number, would be 28.

23. Four fifths of the number would be $21 + 7 = 28$; then $\frac{1}{5}$ is $\frac{1}{4}$ of $28 = 7$, and $\frac{5}{5} = 35$.

25. Since Sarah has 3 cents less than Mary, she has only 5 cents more than Jane. Three times Jane's money is 43 ct. — 8 ct. — 5 ct. = 30 ct.; then Jane's money is $\frac{1}{3}$ of 30 ct. = 10 ct.; Mary's is 10 ct. + 8 ct. = 18 ct.; Sarah's is 10 ct. + 5 ct. = 15 ct.

26. Three times Frank's age = 42 yr. + 3 yr., which is 45 yr.; then Frank's age is $\frac{1}{3}$ of 45 yr. = 15 yr. Mary's age is 2 times 15 yr. less 3 yr. = 27 yr.

27. The ring cost \$5 and the watch \$12 more than the chain; then \$62 — \$12 — \$5 = \$45, which is 3 times the cost of the chain; $\frac{1}{3}$ of \$45 = \$15, the cost of the chain; \$15 + \$5 = \$20, cost of the ring; and \$15 + \$12 = \$27, cost of the watch.

28. One half of $\frac{4}{7}$ is $\frac{2}{7}$. If 30 + 6, or 36, is $\frac{2}{7}$, then $\frac{1}{7}$ is 18, and $\frac{7}{7}$ are 7 times 18 = 126.

29. James has one part; John has two parts + \$3; Frank has three parts + \$3 + \$7; \$55 — \$3 — \$3 — \$7 = \$42, which is 6 times James's money. $\frac{1}{6}$ of \$42 is \$7, James's money; 2 times \$7, + \$3 = \$17, John's money; and 3 times \$7, + \$3, + \$7 = \$31, Frank's.

30. Thomas has 1 part; Joseph has 3 parts less \$2; Paul has 8 parts less \$4 less \$20; then \$20 + \$4 + \$2 + \$22 equal \$48, which is 12 times Thomas's money. $\frac{1}{12}$ of \$48 is \$4, Thomas's money; 3 times \$4, — \$2 = \$10; Joseph's money; 8 times \$4, — \$24 = \$8, Paul's money.

31. The harness cost 1 part; the horse, 1 part + \$50; the buggy, 2 parts + \$50 + \$25; then \$225 — \$50 — \$50 — \$25 = \$100, which is 4 times the cost of the harness. $\frac{1}{4}$ of \$100 is \$25, cost of harness; \$25 + \$50 = \$75, cost of horse; 2 times \$25 + \$75 = \$125, cost of buggy.

LESSON LVII.

2. Both have to pay $\frac{3}{7} + \frac{7}{7} = \frac{10}{7}$; $\frac{1}{7}$ is $\frac{1}{10}$ of \$60 = \$6. John pays 3 times \$6 = \$18; Thomas pays 7 times \$6 = \$42.

3. Four fourths + $\frac{3}{4} = \frac{7}{4}$; $\frac{1}{4}$ is $\frac{1}{7}$ of 56 mi. = 8 mi.; $\frac{4}{4}$ are 32 mi., and $\frac{3}{4}$ are 24 mi., the distance traveled each day, respectively.

4. Since the first, plus $\frac{5}{7}$ of the first, less 8 (that is $\frac{12}{7}$ of the first less 8), = 100, then $\frac{12}{7}$ of the first = 108; $\frac{1}{7}$ is $\frac{1}{12}$ of 108 = 9; $\frac{7}{7}$ = 63, the first; $\frac{5}{7}$ = 45, and 45 less 8 = 37, the second.

5. Four fourths $+ \frac{2}{4} + \frac{3}{4} = \frac{9}{4}$, or 45; $\frac{1}{4}$ is $\frac{1}{9}$ of $45 = 5$. $\frac{4}{4} = 20$, the first part; $\frac{2}{4} = 10$, the second part; $\frac{3}{4} = 15$, the third part.

10. If $\frac{1}{2}$ of the cows $= \frac{2}{7}$ of the sheep, then all of the cows $= \frac{4}{7}$ of the sheep, and $1 + \frac{4}{7} = \frac{11}{7}$ of the sheep; $\frac{1}{7}$ of the sheep is $\frac{1}{11}$ of $55 = 5$; $\frac{7}{7} = 35$, the number of sheep; $\frac{4}{7} = 20$, the number of cows.

11. If $\frac{1}{3}$ of the less $= \frac{2}{9}$ of the greater, $\frac{3}{3}$, or the whole of the less, $= 3$ times $\frac{2}{9}$, which is $\frac{6}{9} = \frac{2}{3}$; then $\frac{3}{3} + \frac{2}{3} = 60$; $\frac{1}{3}$ is $\frac{1}{5}$ of $60 = 12$; $\frac{3}{3} = 36$, the greater number; $\frac{2}{3} = 24$, the smaller number.

12. If $\frac{1}{4}$ of Mary's age $= \frac{1}{3}$ of Sarah's, $\frac{4}{4}$ of Mary's age $= \frac{4}{3}$ of Sarah's; $\frac{4}{3} + \frac{3}{3} = \frac{7}{3}$; $\frac{1}{3}$ of Sarah's age is $\frac{1}{7}$ of $14 = 2$; $\frac{3}{3} = 6$, Sarah's age; $\frac{4}{3} = 8$, Mary's age.

13. If $\frac{2}{3}$ of the first $= \frac{3}{4}$ of the second, $\frac{1}{3}$ is $\frac{1}{2}$ of $\frac{3}{4} = \frac{3}{8}$, and $\frac{3}{3}$, or the whole of the first, $= \frac{9}{8}$ of the second. If the first is $\frac{9}{8}$ of the second, and the second $\frac{8}{8}$, both $= \frac{17}{8}$ of the second. $\frac{1}{8}$ is $\frac{1}{17}$ of $51 = 3$; $\frac{9}{8} = 27$, the first part; $\frac{8}{8} = 24$, the second.

14. If $\frac{2}{3}$ of the apple-trees $= \frac{4}{7}$ of the peach-trees, $\frac{1}{3} = \frac{2}{7}$, and $\frac{3}{3} = \frac{6}{7}$; $\frac{3}{3} = \frac{7}{7}$; then $\frac{7}{7}$ of the peach-trees $+ \frac{6}{7}$ of the peach-trees $= \frac{13}{7}$ of the peach-trees, and $\frac{1}{7}$ is $\frac{1}{13}$ of 65 trees $= 5$ trees. $\frac{7}{7} = 35$, the number of peach-trees; $\frac{6}{7} = 30$, the number of apple-trees.

15. If $\frac{2}{3}$ of A's distance $= \frac{5}{9}$ of B's, then $\frac{1}{3} = \frac{5}{18}$, and $\frac{3}{3} = \frac{15}{18}$, or $\frac{5}{6}$; then A travels $\frac{5}{6}$ as far as B, and both traveled $\frac{5}{6} + \frac{6}{6} = \frac{11}{6}$, or 66 miles. $\frac{1}{6}$ is $\frac{1}{11}$ of 66 mi. $= 6$ mi.; $\frac{6}{6} = 36$ mi., B's distance; $\frac{5}{6} = 30$ mi., A's distance; and 36 mi. $-$ 30 mi. $= 6$ mi., the number of miles B traveled more than A.

16. Let $\frac{12}{12}$ = the apple-trees; $\frac{4}{12}$ = the plum-trees; $\frac{1}{2}$ of $\frac{12}{12}$ + $\frac{1}{4}$ of $\frac{4}{12}$ = $\frac{7}{12}$, the cherry-trees; then $\frac{12}{12}+\frac{4}{12}+\frac{7}{12}=\frac{23}{12}$, or 69 trees. $\frac{1}{12}$ is $\frac{1}{23}$ of 69 = 3; $\frac{12}{12}$ = 36, the number of apple-trees; $\frac{4}{12}$ = 12, the plum-trees; and $\frac{7}{12}$ = 21, the cherry-trees.

17. Five thirds of 12 yr. are 20 yr. If 20 yr. are $\frac{4}{9}$ of both Jane's and Sarah's age, $\frac{1}{9}$ is 5 yr., and $\frac{9}{9}$ are 45 yr. If Jane's age is $\frac{7}{8}$ of Sarah's, then $\frac{8}{8}+\frac{7}{8}=\frac{15}{8}$, and $\frac{15}{8}$ = 45 yr.; $\frac{1}{8}$ = 3 yr., and $\frac{8}{8}$ = 24 yr., Sarah's age; $\frac{7}{8}$ = 21 yr., Jane's age.

18. Three elevenths of 44 are 12; $\frac{4}{5}$ of 30 is 24; 24 is $\frac{4}{9}$ of 54; twice 54 are 108; 12 is contained in 108 nine times.

19. John's money is $\frac{3}{5}$, and Charles's $\frac{5}{5}$, or $\frac{20}{20}$; $\frac{3}{4}$ of $\frac{3}{5}$ = $\frac{9}{20}$, and $\frac{9}{20}$ + \$33 = $\frac{20}{20}$, Charles's money; then $\frac{11}{20}$ = \$33. $\frac{1}{20}$ = \$3, and $\frac{20}{20}$ = \$60, Charles's money; $\frac{3}{5}$ of \$60 = \$36, John's money.

20. Let $\frac{12}{12}$ = the hogs; then $\frac{8}{12}$ = the sheep, and $\frac{6}{12}$ the cows; $\frac{12}{12}+\frac{8}{12}+\frac{6}{12}=\frac{26}{12}$, and $\frac{26}{12}$ = 104; $\frac{1}{12}$ = 4. $\frac{12}{12}$ = 48, the hogs; $\frac{8}{12}$ = 32, the sheep; and $\frac{6}{12}$ = 24, the cows.

22. From noon to midnight is 12 hr. If the time elapsed since noon is $\frac{3}{5}$ of the time to midnight, then it is still $\frac{5}{5}$ to midnight; $\frac{3}{5}+\frac{5}{5}=\frac{8}{5}$, and $\frac{8}{5}$ = 12 hr; $\frac{1}{5}$ is $\frac{1}{8}$ of 12 hr. = $1\frac{1}{2}$ hr.; $\frac{3}{5}$ = $4\frac{1}{2}$ hr. Therefore it is half-past four o'clock, P. M.

23. Since once the time past noon + 3 hr. is $\frac{1}{2}$ the time to midnight, twice the time past noon + 6 hr. = the whole time to midnight; but the time past noon +

the time to midnight is 12 hr.; hence the time past noon, with twice the time past noon + 6 hr. = 12 hr.; hence 3 times the time past noon is 6 hr., and the time past noon is $\frac{1}{3}$ of 6 hr. = 2 hr.

24. Let $\frac{5}{5}$ = the whole time; the time past noon is $\frac{1}{5}$; from midnight to noon is $\frac{4}{5}$; then $\frac{4}{5}$ = 12 hr; $\frac{1}{5}$ = 3 hr. It is 3 o'clock in the afternoon.

25. Let $\frac{4}{4}$ = the whole time; from midnight to noon is $\frac{3}{4}$; the time past noon = $\frac{1}{4}$; $\frac{3}{4}$ = 12 hr.; $\frac{1}{4}$ = 4 hr. It is 4 o'clock in the afternoon.

26. If $\frac{1}{2}$ the time past noon = $\frac{1}{20}$ of the time past midnight, the whole time past noon = $\frac{1}{10}$ the time past midnight; $\frac{10}{10}$ = the whole time; $\frac{1}{10}$ = the time past noon; $\frac{9}{10}$ = the time from midnight to noon, or 12 hr.; $\frac{1}{10}$ = $1\frac{1}{3}$ hr. It is 20 min. past one o'clock P. M.

LESSON LVIII.

2. One + $\frac{2}{3}$ = $\frac{5}{3}$; $\frac{1}{3}$ is $\frac{1}{5}$ of 20 = 4; $\frac{3}{3}$ = 12, the number.

4. Twice the number is $\frac{10}{5}$, and $\frac{10}{5} + \frac{3}{5} = \frac{13}{5}$. If $\frac{13}{5}$ = 52, then $\frac{1}{5}$ is $\frac{1}{13}$ of 52 = 4, and $\frac{5}{5}$ = 20, the number.

5. Twice the number is $\frac{14}{7}$, and $\frac{14}{7}$ less $\frac{4}{7} = \frac{10}{7}$. If $\frac{10}{7}$ = 40, then $\frac{1}{7}$ is $\frac{1}{10}$ of 40, which is 4, and $\frac{7}{7}$ = 28.

6. Let $\frac{5}{5}$ = the number; 3 times $\frac{5}{5}$ less $\frac{3}{5} = \frac{12}{5}$. If $\frac{12}{5}$ = 48, then $\frac{1}{5}$ is $\frac{1}{12}$ of 48 = 4, and $\frac{5}{5}$ = 20.

7. Let $\frac{6}{6}$ = his age; then $\frac{6}{6} + \frac{3}{6} + \frac{4}{6} = \frac{13}{6}$. If 26 = $\frac{13}{6}$, then $\frac{1}{6}$ = 2, and $\frac{6}{6}$ = 12.

8. Her age $=\frac{12}{12}$, and $\frac{12}{12}+\frac{4}{12}+\frac{3}{12}=\frac{19}{12}$; twice her age is $\frac{24}{12}$, and $\frac{24}{12}-\frac{19}{12}=\frac{5}{12}$; ten years $=\frac{5}{12}$, and $\frac{1}{12}=$ 2 years; then $\frac{12}{12}$, her age, is 24 years.

9. Five fifths less $\frac{2}{5}=\frac{3}{5}$; $\frac{3}{5}$ are 30 cents, then $\frac{5}{5}$ are 50 cents.

10. Let $\frac{10}{10}=$ the number; $\frac{10}{10}+\frac{5}{10}+\frac{6}{10}=\frac{21}{10}$; three times the number is $\frac{30}{10}$; $\frac{30}{10}-\frac{21}{10}=\frac{9}{10}$; $27=\frac{9}{10}$; $\frac{1}{10}=$ 3; $\frac{10}{10}=30$.

11. Let $\frac{11}{11}=$ the father's age; $\frac{11}{11}-\frac{3}{11}=\frac{8}{11}$; $\frac{8}{11}=40$ yr.; $\frac{1}{11}=5$ yr.; $\frac{11}{11}=55$ yr.

12. Let $\frac{5}{5}=$ her age; $\frac{5}{5}+\frac{4}{5}=\frac{9}{5}$; three times her age is $\frac{15}{5}$; $\frac{15}{5}-\frac{9}{5}=\frac{6}{5}$; 18 yr. $=\frac{6}{5}$; $\frac{1}{5}=3$ yr.; $\frac{5}{5}=15$ yr.

13. Let $\frac{9}{9}=$ the whole length; then $\frac{9}{9}-\frac{2}{9}=\frac{7}{9}$; $\frac{7}{9}=$ 28 yd.; $\frac{1}{9}=4$ yd.; $\frac{9}{9}=36$ yd.

14. Let $\frac{3}{6}=$ the distance from A to B, and $\frac{6}{6}$ the distance from C to D; $\frac{2}{3}$ of $\frac{3}{6}$ are $\frac{6}{18}=\frac{2}{6}$. $\frac{2}{6}+20=\frac{6}{6}$; then $20=\frac{4}{6}$; $\frac{1}{6}=5$; $\frac{6}{6}=30$, the distance from C to D; $\frac{3}{6}=$ 15, the distance from A to B.

15. Let $\frac{15}{15}=$ my age; $\frac{15}{15}+\frac{5}{15}+\frac{3}{15}=\frac{23}{15}$; $\frac{2}{3}$ of 69 years $=46$ years; $\frac{23}{15}=46$ years; $\frac{1}{15}=2$ years; $\frac{15}{15}=30$ years.

LESSON LIX.

4. As many lots as $\frac{3}{8}$ are contained times in $\frac{8}{8}$, which are $2\frac{2}{3}$.

5. In $2\frac{1}{2}$ days are $\frac{5}{2}$ days; in $\frac{1}{2}$ day he would do $\frac{1}{5}$ of the work; in 1 day he would do $\frac{2}{5}$ of the work.

7. In $3\frac{1}{3}$ days are $\frac{10}{3}$ days. In $\frac{1}{3}$ of a day he would walk $\frac{1}{10}$; in 1 day, $\frac{3}{10}$; in 2 days, $\frac{6}{10}=\frac{3}{5}$.

8. Both do $\frac{1}{2}+\frac{1}{4}$, which are $\frac{3}{4}$.

9. All do the sum of $\frac{1}{2}$, $\frac{1}{4}$ and $\frac{1}{5}$, which is $\frac{19}{20}$.

11. A digs $\frac{1}{6}$ in 1 day; B digs $\frac{1}{12}$ in 1 day; both dig $\frac{3}{12}=\frac{1}{4}$ in 1 day. If they dig $\frac{1}{4}$ in 1 day, it will take 4 days to dig the whole trench.

12. C does $\frac{1}{5}$ in 1 day; B does $\frac{1}{7}$ in 1 day; both do $\frac{12}{35}$ in 1 day. It will take as many days to do it all as $\frac{12}{35}$ are contained times in $\frac{35}{35}$, which are $2\frac{11}{12}$

13. A can do $\frac{1}{2}$ in 1 day, B $\frac{1}{3}$, and C $\frac{1}{6}$; all do in 1 day the sum of $\frac{1}{2}$, $\frac{1}{3}$, and $\frac{1}{6}$, which is $\frac{6}{6}$. Therefore, all three do it in 1 day.

15. Both drink $\frac{1}{12}$ in 1 day; the woman drinks $\frac{1}{30}$ in 1 day; the man drinks $\frac{1}{12}-\frac{1}{30}=\frac{1}{20}$ in 1 day. If he drink $\frac{1}{20}$ in 1 day, he would drink it all in 20 days.

16. All do $\frac{1}{4}$ in 1 day; A and B do $\frac{1}{8}+\frac{1}{12}=\frac{5}{24}$ in 1 day; C does in 1 day $\frac{1}{4}-\frac{5}{24}=\frac{1}{24}$. Therefore, C can reap it all in 24 days.

17. Both do in 1 day $\frac{1}{2}+\frac{1}{3}=\frac{5}{6}$. If they do $\frac{5}{6}$ in 1 day, they would do it all in $1\frac{1}{5}$ days.

18. A digs $\frac{2}{5}$ in 1 day, and B digs $\frac{3}{10}$ in 1 day; both dig $\frac{2}{5}+\frac{3}{10}=\frac{7}{10}$ in 1 day. Therefore, if they dig $\frac{7}{10}$ in 1 day, they would dig it all in $1\frac{3}{7}$ days.

19. C reaps $\frac{1}{5}$ in 1 day; D reaps $\frac{3}{10}$ in 1 day; both reap in 1 day $\frac{1}{5}+\frac{3}{10}=\frac{5}{10}=\frac{1}{2}$. If they reap $\frac{1}{2}$ in 1 day, they would reap the whole in 2 days.

LESSON LX.

8. The ratio of 21 to 7 is 3; $36 \div 3 = 12$, the number.

9. The ratio of 20 to 2 is 10; $10 - 5 = 5$, 5 is $\frac{1}{4}$ of 20, and 20 is the ratio of 40 to 2.

10. The ratio of 18 to 2 is $9, + 3 = 12, + 7 = 19$; and 19 is the ratio of 38 to 2.

11. The ratio of 27 to 9 is $3, + 5 = 8$; and 8 is the ratio of 20 to $2\frac{1}{2}$.

13. Five $+ 7 = 12$; $\frac{5}{12}$ of $48 = 20$, the first part; $\frac{7}{12}$ of $48 = 28$, the second part.

18. Once the number + 3 times the number = 4 times the number, or 48; and $\frac{1}{4}$ of 48 is 12, the number.

19. One $+ 4 = 5$; the first is $\frac{1}{5}$ of 25 yd. = 5 yd.; the second is $\frac{4}{5} = 20$ yd.

22. The first has $\frac{2}{3}$ of $7\frac{1}{2}$ doz. = 5 doz.; the second has $\frac{1}{3}$ of $7\frac{1}{2}$ doz. = $2\frac{1}{2}$ doz.

23. A paid $\frac{25}{40} = \frac{5}{8}$ of the cost, and B paid $\frac{15}{40} = \frac{3}{8}$ of the cost. A should receive $\frac{5}{8}$ of $56 = $35, and B should receive $\frac{3}{8}$ of $56 = $21.

24. Three $+ 2 = 5$. C's loss was $\frac{3}{5}$ of $30 = $18; D's loss was $\frac{2}{5}$ of $30 = $12.

LESSON LXI.

3. Three thirds $+ \frac{4}{3} = \frac{7}{3}$. A has $\frac{3}{7}$ of 14 ct. $= 6$ ct.; B has $\frac{4}{7}$ of 14 ct. $= 8$ ct.

5. In $2\frac{1}{2}$ are $\frac{5}{2}$, and $4\frac{1}{2} = \frac{9}{2}$; $\frac{5}{2} + \frac{9}{2} = \frac{14}{2}$. The first would receive $\frac{5}{14}$ of \$28 $=$ \$10; the second would receive $\frac{9}{14}$ of \$28 $=$ \$18.

6. Three thirds $+ \frac{5}{3} = \frac{8}{3}$. William's age is $\frac{5}{8}$ of 32 yr. $= 20$ yr.; Frank's age is $\frac{3}{8}$ of 32 yr. $= 12$ yr.

7. Three thirds $+ \frac{7}{3} = \frac{10}{3}$. $\frac{7}{10}$ of 30 apples $= 21$ apples, the number of sound ones; $\frac{3}{10}$ of 30 apples $= 9$ apples, the number not sound.

8. Four fifths $+ \frac{5}{5} = \frac{9}{5}$. One built $\frac{4}{9}$ of 27 ft. $= 12$ ft.; the other, $\frac{5}{9} = 15$ ft.

10. One $+ 2 + 3 + 4 = 10$. The first part is $\frac{1}{10}$ of $70 = 7$; the second, $\frac{2}{10}$ of $70 = 14$; the third, $\frac{3}{10}$ of $70 = 21$; the fourth, $\frac{4}{10}$ of $70 = 28$.

11. One half $= \frac{6}{12}$; $\frac{1}{3} = \frac{4}{12}$; $\frac{1}{4} = \frac{3}{12}$; and $\frac{6}{12} + \frac{4}{12} + \frac{3}{12} = \frac{13}{12}$. The first is $\frac{6}{13}$ of $39 = 18$; the second, $\frac{4}{13} = 12$; the third is $\frac{3}{13} = 9$.

12. All had 3 ct. $+ 4$ ct. $+ 5$ ct. $= 12$ ct. William's share, $\frac{3}{12}$ of 36, $= 9$; Thomas's, $\frac{4}{12}$ of 36, $= 12$; John's, $\frac{5}{12}$ of 36, $= 15$.

13. The whole loss was \$864 — \$500 $=$ \$364; $\frac{1}{8}$ of \$364 is \$$45\frac{1}{2}$, A's loss; $\frac{1}{4}$ of \$364 is \$91, B's loss; $\frac{5}{8}$ of \$364 are \$$227\frac{1}{2}$, C's loss.

14. A has $\frac{4}{4}$, B $\frac{2}{4}$, and C $\frac{1}{4}$; all have $\frac{4}{4}+\frac{2}{4}+\frac{1}{4}=\frac{7}{4}$. A has $\frac{4}{7}$ of \$42 = \$24; B has $\frac{2}{4}$ of \$24 = \$12; and C has $\frac{1}{4}$ of \$24 = \$6.

15. Four + 3 + 2 = 9. A has $\frac{4}{9}$, B $\frac{3}{9}$, and C $\frac{2}{9}$, or 20, 15, and 10, respectively.

16. One + 3 + 6 = 10. $\frac{1}{10}$ of 60 = 6, the horses; $\frac{3}{10}$ of 60 = 18, the cows; $\frac{6}{10}$ of 60 = 36, the sheep.

17. One + 2 + 3 = 6. A has $\frac{1}{6}$ of 42 = 7; B, $\frac{2}{6}$ = 14; C, $\frac{3}{6}$ = 21.

18. One + 2 + 4 = 7. Emma has $\frac{1}{7}$ of 35 = 5; Agnes has $\frac{2}{7}$ of 35 = 10; Sarah has $\frac{4}{7}$, = 20.

LESSON LXII.

2. It will take 15 times 8 men = 120 men to do the work in 1 day; to do it in 12 days it will take $\frac{1}{12}$ of 120 men = 10 men.

4. One will fill it in 9 times $2\frac{1}{2}$ hr. = $22\frac{1}{2}$ hr.; then 5 pipes will fill it in $\frac{1}{5}$ of $22\frac{1}{2}$ hr. = $4\frac{1}{2}$ hr

6. Fifteen ct. = \$$\frac{3}{20}$; 80 times \$$\frac{3}{20}$ = \$$\frac{240}{20}$ = \$12.

7. Sixty ct. = \$$\frac{3}{5}$; 80 times \$$\frac{3}{5}$ = \$$\frac{240}{5}$ = \$48.

10. It will make 3 times 20 = 60 one cent loaves. $\frac{1}{4}$ of 60 = 15 four cent loaves; $\frac{1}{5}$ of 60 = 12 five cent loaves.

11. A loaf will weigh 3 times 8 oz. = 24 oz., when flour is \$1 a barrel; it will weigh $\frac{1}{4}$ of 24 oz. = 6 oz., when flour is \$4 a barrel.

12. Six times 10 oz. $=$ 60 oz.; $\frac{1}{5}$ of 60 oz. $=$ 12 oz.

13. In \$$5\frac{1}{3}$ are \$$\frac{16}{3}$; $\frac{16}{3}$ times 7 oz. $= \frac{112}{3}$ oz.; \$$4\frac{2}{3} =$ \$$\frac{14}{3}$. When flour is worth \$$4\frac{2}{3}$ a barrel, a loaf will weigh as many oz. as $\frac{14}{3}$ are contained times in $\frac{112}{3} = 8$.

14. It will take 5 times 5 men $=$ 25 men to do the same work in $\frac{1}{5}$ of the time; to do twice as much will take 2 times 25 men $=$ 50 men.

15. Six men will do $\frac{1}{2}$ of it in $\frac{1}{2}$ of 5 days $= 2\frac{1}{2}$ days; one man will do the other $\frac{1}{2}$ in 6 times $2\frac{1}{2}$ days $=$ 15 days. $6 + 3 = 9$; 9 men will do it in $\frac{1}{9}$ of 15 days $= 1\frac{2}{3}$ days. Therefore, the whole time is $2\frac{1}{2}$ days $+ 1\frac{2}{3}$ days $=$ $4\frac{1}{6}$ days.

16. Seven men will do $\frac{1}{2}$ of the work in 2 days; one man will do the other half in 7 times 2 days $=$ 14 days. $7 - 3 = 4$; 4 men will do it in $\frac{1}{4}$ of 14 days $= 3\frac{1}{2}$ days. Therefore, it will take 2 days $+ 3\frac{1}{2}$ days $= 5\frac{1}{2}$ days to do the whole work.

18. One man would spend $\frac{1}{6}$ of \$36 $=$ \$6 in 8 days; in 1 day he would spend $\frac{1}{8}$ of \$6 $=$ \$$\frac{6}{8}$. 5 persons would spend 5 times \$$\frac{6}{8} =$ \$$\frac{30}{8}$ in 1 day; in 12 days, 12 times \$$\frac{30}{8} =$ \$$\frac{360}{8} =$ \$45.

19. One third of 12 rd. is 4 rd.; $\frac{1}{8}$ of 4 rd. is $\frac{4}{8}$ rd. $= \frac{1}{2}$ rd; 5 times $\frac{1}{2}$ rd. $= \frac{5}{2}$ rd.; 3 times $\frac{5}{2}$ rd. $= \frac{15}{2}$ rd. $= 7\frac{1}{2}$ rd.

20. One sixth of 36 bu. $=$ 6 bu.; $\frac{1}{10}$ of 6 bu. $= \frac{3}{5}$ bu.; 5 times $\frac{3}{5}$ bu. $= \frac{15}{5}$ bu.; 9 times $\frac{15}{5}$ bu. $= \frac{135}{5}$ bu. $=$ 27 bu.

21. One man will eat 2 B. in 5 times 6 days = 30 days; one man will eat 8 B. in 4 times 30 days = 120 days; 12 men will eat 8 B. in $\frac{1}{12}$ of 120 days = 10 days.

22. One person would spend $\frac{1}{8}$ of \$400 = \$50 in 5 months; in 1 month he would spend $\frac{1}{5}$ of \$50 = \$10; in 8 mo. 1 person would spend 8 times \$10 = \$80, and 11 persons would spend 11 times \$80 = \$880.

23. One ox can be kept on $\frac{1}{10}$ of 5 T. = $\frac{1}{2}$ T. for 3 mo.; one ox can be kept 1 mo. on $\frac{1}{3}$ of $\frac{1}{2}$ T. = $\frac{1}{6}$ T.; for 5 mo. on 5 times $\frac{1}{6}$ T. = $\frac{5}{6}$ T. 15 T. = $\frac{90}{6}$ T., and $\frac{90}{6}$ T. will keep as many oxen as $\frac{5}{6}$ are contained times in $\frac{90}{6}$ = 18. If 7 sheep eat as much as an ox, it will keep 18 times 7 sheep = 126 sheep.

LESSON LXIII.

2. One horse eats as much as $1\frac{1}{2}$ cows, and 14 horses eat as much as 21 cows; 15 cows + 21 cows = 36 cows. A pays $\frac{21}{36} = \frac{7}{12}$ of \$60 = \$35; B pays $\frac{15}{36} = \frac{5}{12}$ of \$60 = \$25.

3. B's 120 sheep = 6 horses; his 15 oxen = 10 horses; then B has the same as 6 horses + 10 horses = 16 horses. Both have 16 horses + 8 horses = 24 horses. A pays $\frac{8}{24}$, or $\frac{1}{3}$, of \$72 = \$24; B pays $\frac{2}{3}$ of \$72 = \$48.

5. C's \$50 for 4 mo. = \$200 for 1 mo.; D's \$60 for 5 mo. = \$300 for 1 mo.; \$200 + \$300 = \$500. C has $\frac{2}{5}$ of \$45 = \$18; D has $\frac{3}{5}$ of \$45 = \$27.

6. Three men for 4 days = 12 men 1 day; 5 men 3 days = 15 men 1 day; 12 men + 15 men = 27 men. A receives $\frac{4}{9}$ of \$81 = \$36; B receives $\frac{5}{9}$ of \$81 = \$45.

7. A's \$2 for 5 mo. = \$10 for 1 mo.; B's \$3 for 4 mo. = \$12 for 1 mo.; \$10 + \$12 = \$22. A receives $\frac{5}{11}$ of \$55 = \$25; B, $\frac{6}{11}$ = \$30.

8. E's 4 horses = 6 cows; 6 cows for 5 mo. = 30 cows 1 mo.; F's 10 cows for 6 mo. = 60 cows 1 mo.; 30 cows + 60 cows = 90 cows. E pays $\frac{1}{3}$ of \$27 = \$9; F pays $\frac{2}{3}$ of \$27 = \$18.

9. The net gain is \$300 — \$150 = \$150; \$600 + \$900 = \$1500. M has $\frac{2}{5}$ of \$150 = \$60; N has $\frac{3}{5}$ of \$150 = \$90.

10. C's capital = \$600 for 12 mo., or \$7200 for 1 mo.; D's = \$600 for 8 mo., or \$4800 for 1 mo.; \$7200 + \$4800 = \$12000. C has $\frac{72}{120} = \frac{3}{5}$ of \$250 = \$150; D has $\frac{48}{120} = \frac{2}{5}$ of \$250 = \$100.

11. E had \$1000 for 12 mo. = \$12000 for 1 mo.; F had \$3000 for 12 mo. = \$36000 for 1 mo., less \$1000 for 4 mo. = \$4000 for 1 mo., and \$36000 — \$4000 = \$32000; \$32000 + \$12000 = \$44000. E has $\frac{3}{11}$ of \$770 = \$210; F has $\frac{8}{11}$ of \$770 = \$560.

12. \$240 — \$20 = \$220; B's share is $\frac{1}{2}$ of \$220 = \$110; A's share is \$110 + \$20 = \$130. B has $\frac{110}{240} = \frac{11}{24}$ of the gain, and also $\frac{11}{24}$ of the capital, \$2400, = \$1100; A has $\frac{13}{24}$ of the gain, and also $\frac{13}{24}$ of \$2400, = \$1300.

13. Since D's time was only $\frac{3}{4}$ of C's, he must have had $\frac{1}{4}$ more capital than C; then $\frac{4}{4}$ = D's capital, and $\frac{3}{4}$ = C's capital; $\frac{4}{4} + \frac{3}{4} = \frac{7}{4}$, and $\frac{7}{4}$ = \$980. $\frac{1}{4} = \frac{1}{7}$ of \$980 = \$140; $\frac{4}{4}$ = \$560, D's capital; $\frac{3}{4}$ = \$420, C's capital.

14. A's gain per mo. was $\frac{1}{10}$ of \$70 = \$7; B's gain per mo. was $\frac{1}{8}$ of \$80 = \$10. Both gained \$7 + \$10 = \$17 per mo. If A has $\frac{7}{17}$ of the gain, his capital is $\frac{7}{17}$ of \$1700 = \$700; B's capital is $\frac{10}{17}$ of \$1700 = \$1000.

15. The ratio of their stock was as 2 to 3, and of the time as 10 to 12; $2 \times 10 = 20$; $3 \times 12 = 36$; $20 + 36 = 56$. E's gain was $\frac{5}{14}$ of \$840 = \$300; F's gain was $\frac{9}{14}$ of \$840 = \$540.

LESSON LXVII.

8. $\frac{30}{45} = \frac{2}{3} = 66\frac{2}{3}$ per cent.

9. He gains $\frac{5}{25} = \frac{1}{5}$, and $\frac{1}{5} = 20$ per cent.

14. He gains \$24 — \$15 = \$9; \$9 = $\frac{3}{5}$ of the cost, or 60 per cent.

15. In 5 gal. are 20 qt. He lost $\frac{6}{20} = \frac{3}{10} = 30$ per cent.

16. Six cents = $\frac{6}{5}$ of the cost; $\frac{5}{5} = 5$ cents, the cost; 8 ct. — 5 ct. = 3 ct.; $\frac{3}{5} = 60$ per cent.

17. He paid $\frac{1}{6}$ of \$3 = 50 cents for 1 yd.; he sold 1 yd. for $\frac{1}{5}$ of \$4 = 80 ct.; 80 ct. — 50 ct. = 30 ct.; $\frac{30}{50} = \frac{3}{5}$ = 60 per cent.

18. Eight ct. = $\frac{4}{5}$ of the cost; $\frac{5}{5} = 10$ ct., the cost. $\frac{1}{3}$ of 25 ct. = $8\frac{1}{3}$ ct.; $10 - 8\frac{1}{3} = 1\frac{2}{3} = \frac{1\frac{2}{3}}{10} = \frac{5}{30} = \frac{1}{6} = 16\frac{2}{3}$ per cent loss. $\frac{1}{2}$ of 25 ct. = $12\frac{1}{2} - 10 = 2\frac{1}{2}$; $\frac{2\frac{1}{2}}{10} = \frac{5}{20} = \frac{1}{4}$; $\frac{1}{4} = 25$ per cent gain.

19. One lemon cost $\frac{1}{2}$ of 3 ct. = $1\frac{1}{2}$ ct. He sold 1 lemon for $\frac{1}{3}$ of 2 ct. = $\frac{2}{3}$ ct. $1\frac{1}{2} = \frac{9}{6}$; $\frac{2}{3} = \frac{4}{6}$; $\frac{9}{6} - \frac{4}{6} = \frac{5}{6}$; $\frac{5}{6}$ are $\frac{5}{9}$ of $\frac{9}{6}$, and $\frac{5}{9} = 55\frac{5}{9}$ per cent.

LESSON LXVIII.

2. \$26 = 130 per cent of the cost, or $\frac{13}{10}$; $\frac{1}{10} = \frac{1}{13}$ of \$26 = \$2; $\frac{10}{10}$ = \$20, the cost.

3. Fourteen cents = $\frac{14}{10}$ of the cost; $\frac{1}{10}$ is $\frac{1}{14}$ of 14 ct. = 1 ct.; $\frac{10}{10}$ = 10 ct., the cost.

4. \$81 = $\frac{9}{8}$ of the cost; $\frac{1}{8}$ = \$9; $\frac{8}{8}$ = \$72.

5. \$63 = $\frac{9}{10}$ of the cost; $\frac{1}{10}$ = \$7; $\frac{10}{10}$ = \$70.

6. \$21 = $\frac{7}{4}$ of the cost; $\frac{1}{4}$ = \$3; $\frac{4}{4}$ = \$12.

7. Forty ct. = $\frac{4}{3}$ of the cost; $\frac{1}{3}$ = 10 ct.; $\frac{3}{3}$ = 30 ct. Each orange cost $\frac{1}{10}$ of 30 ct. = 3 ct.

8. \$10 = $\frac{5}{4}$ of the cost; $\frac{1}{4}$ = \$2; $\frac{4}{4}$ = \$8, the cost; \$12 — \$8 = \$4. He would gain $\frac{4}{8}$ = 50 per cent.

9. Seven ct. = $\frac{7}{8}$ of the cost; $\frac{1}{8}$ = 1 ct.; $\frac{8}{8}$ = 8 ct., the cost. 8 ct. — 6 ct. = 2 ct.; and 2 ct. are $\frac{1}{4}$ of the cost = 25 per cent loss.

10. \$35 = $\frac{5}{6}$ of the cost; $\frac{1}{6}$ = \$7; $\frac{6}{6}$ = \$42, the cost. \$63 — \$42 = 21; \$21 is $\frac{1}{2}$ of the cost = 50 per cent gain.

11. \$18 = $\frac{6}{5}$ of its value; $\frac{5}{5}$ = \$15, its value; 10 per cent of \$15 is $\frac{1}{10}$ of \$15 = \$1.50; \$18 — \$15 = \$3. He lost \$3 + \$1$\frac{1}{2}$ = \$4$\frac{1}{2}$.

12. \$60 = $\frac{6}{5}$ of the cost; $\frac{5}{5}$ = \$50, the cost. A gained \$60 — \$50 = \$10; B lost 20 per cent, or $\frac{1}{5}$ of \$60 = \$12. B lost \$12 — \$10 = \$2 more than A gained.

13. \$30 $= \frac{5}{4}$ of the cost of the first; $\frac{4}{4} =$ \$24, the cost. The gain was \$30 — \$24 = \$6. \$30 $= \frac{3}{4}$ of the cost of the second; $\frac{4}{4} =$ \$40, the cost. The loss on the watch was \$40 — \$30 = \$10; loss by sale \$10 — \$6 = \$4.

14. One apple sold for $\frac{1}{4}$ of 3 ct. $= \frac{3}{4}$ ct.; $\frac{3}{4}$ ct. $= 150\%$, or $\frac{3}{2}\%$ of the cost. $\frac{1}{2} = \frac{1}{4}$ ct.; $\frac{2}{2} = \frac{2}{4}$ ct. $= \frac{1}{2}$ ct., the cost. 5 apples for 4 ct. $= \frac{4}{5}$ ct. for 1 apple. Gain $\frac{4}{5} - \frac{1}{2} = \frac{3}{10}$ and $\frac{3}{10} = \frac{3}{5}$ of $\frac{5}{10}$, or 60 per cent.

15. One lemon sold for $\frac{4}{5}$ ct.; $\frac{4}{5}$ ct. $= \frac{4}{5}$ per cent of the cost; 1 ct. = the cost; 6 for 5 ct. $= \frac{5}{6}$ ct. for 1; 1 ct. — $\frac{5}{6}$ ct. $= \frac{1}{6}$ ct.; $\frac{1}{6}$ of 1 ct. $= 16\frac{2}{3}$ per cent.

16. Ten per cent of 60 = 6; $\frac{2}{3}$ of 6 = 4; 4 is $\frac{1}{2}$ of 8; 8 $= \frac{1}{5}$, or 20 per cent, of 40.

17. Fifty per cent of 120 = 60; $\frac{3}{5}$ of 60 = 36; $\frac{1}{2}$ of 36 = 18; 18 is 10 less than 28, and 28 is 20 per cent, or $\frac{1}{5}$, of 140.

18. Sixty per cent of 10 is 6; $\frac{2}{3}$ of 6 = 4; $\frac{1}{4}$ of 4 = 1. 1 is 5 less than 6, and 6 is 50 per cent, or $\frac{1}{2}$, of 12.

19. Seventy-five per cent of 15 $= \frac{45}{4}$; $\frac{2}{5}$ of $\frac{45}{4} = \frac{18}{4}$; $\frac{3}{4}$ of $\frac{18}{4} = \frac{27}{8}$; $\frac{27}{8}$ are $1\frac{3}{5}$ more than $\frac{71}{40}$, and $\frac{71}{40}$ are 50 per cent, or $\frac{1}{2}$, of $\frac{142}{40} = 3\frac{11}{20}$.

20. Twenty-five per cent of 4 is 1; $\frac{2}{3}$ of 1 $= \frac{2}{3}$; $\frac{3}{2}$ times $\frac{2}{3} = \frac{3}{3} = 1$; 1 is 25 per cent, or $\frac{1}{4}$, of 4; and 4 is $\frac{1}{2}$ of 8.

LESSON LXIX.

1. One fifth of \$3 is 60 ct.; \$3 — 60 ct. = \$2.40.

2. One fifth of \$125 = \$25; \$125 — \$25 = \$100.

3. One sixth of \$840 = \$140; \$840 — \$140 = \$700.

4. Twenty % = $\frac{1}{5}$; $\frac{1}{5}$ of \$500 = \$100; \$500 — \$100 = \$400. 5% = $\frac{1}{20}$; $\frac{1}{20}$ of \$400 = \$20; \$400 — \$20 = \$380.

5. One sixth of \$1200 = \$200; \$1200 — \$200 = \$1000. 5% = $\frac{1}{20}$; $\frac{1}{20}$ of \$1000 = \$50; \$1000 — \$50 = \$950.

6. \$4.80 = $\frac{4}{5}$ of the retail price; $\frac{1}{5}$ = \$1.20; $\frac{5}{5}$ = \$6.

7. \$720 = $\frac{2}{3}$ of the retail price; $\frac{1}{3}$ = \$360; $\frac{3}{3}$ = \$1080.

8. One hundred per cent less 5 per cent = 95%, or $\frac{19}{20}$; $\frac{19}{20}$ = \$133; $\frac{1}{20}$ = \$7; $\frac{20}{20}$ = \$140. \$140 is $\frac{4}{5}$ of the list price; $\frac{1}{5}$ = \$35; $\frac{5}{5}$ = \$175.

9. \$399 = $\frac{19}{20}$ of the remainder after $\frac{1}{8}$ per cent discount; $\frac{1}{20}$ = \$21; $\frac{20}{20}$ = \$420. \$420 = $\frac{7}{8}$ of the list price; $\frac{1}{8}$ = \$60; $\frac{8}{8}$ = \$480.

10. \$50 less $\frac{1}{5}$ = \$40; \$40 less $\frac{1}{20}$ = \$38; \$20 less $\frac{1}{10}$ = \$18; \$18 less $\frac{1}{20}$ = \$17.10; \$38 + \$17.10 = \$55.10.

LESSON LXX.

1. His commission was $2\frac{1}{2}$ per cent, or $\frac{1}{40}$, of \$4000, = \$100.

2. One twentieth of \$560 = \$28.

3. The corn cost 1000 times 50 ct. = \$500; $\frac{1}{40}$ of \$500 = \$12.50.

4. Five times \$300 = \$1500; $\frac{1}{20}$ of \$1500 = \$75. The owner receives \$1500 — \$75 = \$1425

5. The wheat sells for 800 times \$1.25 = \$1000; $\frac{1}{50}$ of \$1000 = \$20, the commission; \$1000 — \$20 = \$980.

6. \$100 = $\frac{1}{20}$ of \$2000.

7. \$60 = $\frac{1}{10}$; $\frac{10}{10}$ = \$600.

8. The commission was $\frac{1}{20}$ of \$1000 = \$50; \$1000 — \$50 = \$950. He bought as many shares as \$50 are contained times in \$950 = 19.

LESSON LXXI.

1. One per cent of \$2000 = \$20; $\frac{1}{2}\%$ = $\frac{1}{2}$ of \$20 = \$10

2. One half of \$3000 = \$1500; $\frac{1}{50}$ of \$1500 = \$30.

3. Two thirds of \$2400 = \$1600; 1% of \$1600 is \$16, and \$16 + \$1.50 = \$17.50.

4. \$2500 + \$1500 = \$4000; 1% of \$4000 is \$40; $\frac{3}{4}\%$ is $\frac{3}{4}$ of \$40 = \$30.

5. Two thirds of \$1800 are \$1200; $\frac{1}{100}$ of \$1200 is \$12; $\frac{2}{3}$ of \$1200 are \$800; 1% of \$800 is $\frac{1}{100}$ of \$800 = \$8; $\frac{1}{2}\%$ is $\frac{1}{2}$ of \$8 = \$4; and \$12 + \$4 + \$1 = \$17.

6. One half of \$2600 = \$1300; $\frac{1}{50}$ of \$1300 = \$26; $\frac{2}{3}$ of \$1500 are \$1000; $\frac{1}{100}$ of \$1000 is \$10; and \$26 + \$10 + \$1.50 = \$37.50.

LESSON LXXII.

4. The interest on \$1 for 1 year is 8 ct.; for 3 yr. 3 times 8 ct. = 24 ct.; for \$20 it is 20 times 24 ct. = \$4.80.

5. The interest on $1 for 6 yr. at 4% is 24 ct.; and for $25 it is 25 times 24 ct. = $6.

6. The interest for $1 is 20 ct.; for $40 it is $8.

7. $9. (8.) $8.40.

9. $9. (10.) $36.

LESSON LXXIII.

2. Four mo. = $\frac{1}{3}$ of a year. The interest on $1 for 4 mo. is $\frac{1}{3}$ of 5 ct. = $1\frac{2}{3}$ ct.; on $60 it is 60 times $1\frac{2}{3}$ ct. = $1.

3. The interest on $1 for 7 mo. at 6% is $3\frac{1}{2}$ ct.; on $80 it is $2.80.

4. On $1 it is $\frac{3}{4}$ of 8 ct. = 6 ct.; on $40 it is $2.40.

5. Two thirds of 9 ct. = 6 ct.; 75 times 6 ct. = $4.50.

7. The interest for 1 year is 180 times 4 ct. = $7.20; for 1 mo. $\frac{1}{12}$ of $7.20 = $0.60; for 10 mo. 10 times 60 ct. = $6; for 10 days $\frac{1}{3}$ of 60 ct. = 20 ct.; for 10 mo. 10 da., $6.20.

9. One year's interest = $12; 1 mo. interest = $1; for 4 mo. $4; for 24 days $\frac{4}{5}$ of $1 = 80 ct.; for 4 mo. 24 da., $4.80.

10. The int. for 1 yr. is $24; for 1 mo. $2; for 9 mo. $18; for $18 $\frac{3}{5}$ of $2 = $1.20; for 9 mo. 18 da. it is $19.20.

11. $8.45.

12. The int. for 1 yr. is \$5.76; for 1 mo. 48 ct.; for 8 mo. \$3.84; for 25 da. $\frac{5}{6}$ of 48 ct. = 40 ct.; for 8 mo. 25 da., \$3.84 + \$0.40 = \$4.24.

13. \$3.20. (14.) \$6.75. (15.) \$3.80.

16. The int. for 1 year is \$1; for 3 yr. \$3. The amount is \$25 + \$3 = \$28.

17. \$44. (18.) \$68.20. (19.) \$32.80.

20. \$56.80. (21.) \$99.12.

LESSON LXXIV.

2. The int. for \$1 for 3 yr. at 4% is 12 ct. It will take as many dollars to acquire \$6 int. as 12 ct. are contained times in 600 ct., which are 50. *Ans.* \$50.

3. \$60. (4.) \$75. (5.) \$140.

6. \$240. (7.) \$350.

8. As many dollars as 5 ct. are contained times in \$200, which are 4000. *Ans.* \$4000.

LESSON LXXV.

2. The amount of \$1 for 3 yr. at 6% is \$1.18. It will take as many dollars to amount to \$236 as \$1.18 is contained times in \$236, which are 200.

3. \$500. (4.) \$250.

5. \$300. (6.) \$25.

7. The amount of $1 for 2 yr. 6 mo. at 8% is $1.20. It will take as many dollars to amount to $60 as $1.20 is contained times in $60, which are 50. If $50 = $\frac{2}{5}$ of the principal, $\frac{5}{5}$, or the whole, = $125.

LESSON LXXVI.

2. The int. on $40 for 1 yr. at 5% is $2. To gain $8, it will take 4 yr.

3. 2 yr. 6 mo. (4.) 2 yr. 8 mo.

5. $3\frac{3}{7}$ years. (6.) $6\frac{2}{3}$ yr. = 6 yr. 8 mo.

9. Any principal to treble itself must gain 200%. At 5% it will take as many years as 5 is contained times in 200 = 40.

LESSON LXXVII.

2. At 1 per cent the int. on $50 for 5 yr. is $2.50. To amount to $20, the rate will be as many times 1% as $2.50 are contained times in $20, which are 8. *Ans.* 8 per cent.

3. Int. at 1% = $2.25; $11.25 ÷ $2.25 = 5. *Ans.* 5%.

4. 7%.

5. Int. at 1% = $6.75; $54.00 ÷ $6.75 = 8. *Ans.* 8%.

6. Int. at 1% = $8; $56 ÷ $8 = 7. *Ans.* 7%.

7. $240 less $200 = $40, the interest; int. at 1% = $8; $40 ÷ $8 = 5. *Ans.* 5%.

8. \$183 less \$150 = \$33, the int.; int. at 1% = \$5.50; \$33 ÷ \$5.50 = 6. *Ans.* 6%.

LESSON LXXVIII.

2. The amount of \$1, for the given time and rate, is \$1.30 = $\frac{13}{10}$. \$520 = $\frac{13}{10}$; $\frac{1}{10}$ = \$40; $\frac{3}{10}$ = \$120, the discount; $\frac{10}{10}$ = \$400, the present worth.

3. \$30 = $\frac{6}{5}$; $\frac{5}{5}$ = \$25, present worth; \$30 — \$25 = \$5, discount.

4. Present worth, \$500; discount, \$250.

5. \$345 = $\frac{23}{20}$; $\frac{1}{20}$ = \$15; $\frac{20}{20}$ = \$300. \$345 — \$300 = \$45, discount.

6. \$96. (7.) \$4. (8.) \$50. (9.) \$44.

10. Int. for 6 yr. 8 mo. = 40 ct.; amt. of \$1 = \$1.40 = $\frac{7}{5}$. \$77 = $\frac{7}{5}$; $\frac{1}{5}$ = \$11; $\frac{5}{5}$ = \$55, present worth.

11. Amt. of \$1 for 3 yr. 6 mo. at 7% is \$1.24$\frac{1}{2}$ = $\frac{1245}{1000}$ = $\frac{249}{200}$; $\frac{1}{200}$ is $\frac{1}{249}$ of \$1000 = \$4$\frac{4}{249}$; $\frac{200}{200}$ = \$803$\frac{53}{249}$; \$1000 — \$803$\frac{53}{249}$ = \$196$\frac{196}{249}$, discount.

12. Amt. of \$1 is \$1.22 = $\frac{61}{50}$; \$900 = $\frac{61}{50}$; $\frac{1}{50}$ is $\frac{1}{61}$ of \$900 = \$14$\frac{46}{61}$; $\frac{50}{50}$ = \$737$\frac{43}{61}$; \$900 — \$737$\frac{43}{61}$ = \$162$\frac{18}{61}$.

LESSON LXXIX.

1. For 4 yr. 2 mo. 25%; $\frac{25}{100} = \frac{1}{4}$.

2. For 5 yr. 25% int.; 100% + 25% = 125%, amt.; $\frac{25}{125} = \frac{1}{5}$.

3. For 1 yr. $\frac{1}{2}$ of $\frac{1}{5}=\frac{1}{10}$; $\frac{1}{10}$ of 100% is 10%.

4. Two yr. 6 mo. = 30 mo.; $\frac{1}{30}$ of $\frac{1}{4}=\frac{1}{120}$ for 1 mo.; 12 times $\frac{1}{120}=\frac{1}{10}$; $\frac{1}{10}$ of 100% = 10%.

5. The int. at 10% = $\frac{1}{10}$ of the principal in 1 yr.; to equal $\frac{3}{5}$, or $\frac{6}{10}$, it will take 6 yr.

6. The yearly interest is $\frac{1}{3}$ of $\frac{9}{25}=\frac{3}{25}$; $\frac{3}{25}=\frac{12}{100}=12\%$.

7. The interest for 2 yr. is 5 times $\frac{4}{25}=\frac{20}{25}=\frac{4}{5}$; for 1 yr. $\frac{1}{2}$ of $\frac{4}{5}=\frac{2}{5}$; $\frac{2}{5}=\frac{40}{100}=40\%$.

8. Five eighths of the interest for 1 yr. = $\frac{3}{80}$ of the principal. If $\frac{3}{80}$ are $\frac{5}{8}$, then $\frac{1}{8}$ is $\frac{1}{5}$ of $\frac{3}{80}=\frac{3}{400}$, and $\frac{8}{8}=\frac{24}{400}=\frac{6}{100}=6\%$.

9. The int. for 4 mo. is $\frac{1}{50}$ of the principal; for 12 mo. 3 times $\frac{1}{50}=\frac{3}{50}=\frac{6}{100}=6\%$. Int. for \$1 for 1 yr. 4 mo. = 8 ct.; for \$200 it is 200 times 8 ct. = \$16

10. One yr. 4 mo. = 16 mo., or 4 times 4 mo.; $\frac{1}{4}$ of $\frac{3}{25}=\frac{3}{100}$; $\frac{3}{4}=\frac{9}{100}=9\%$. The interest of \$100 for 1 yr. 8 mo. 12 da., at 9%, is \$15.30.

11. In 2 times 4 yr. = 8 yr.

12. In $3\frac{1}{3}$ are $\frac{10}{3}$; $\frac{1}{3}$ is $\frac{1}{10}$ of \$40 = \$4; $\frac{3}{3}$ = \$12; \$12, the int. for 1 yr. = 5%, or $\frac{1}{20}$ of the principal; the principal = $\frac{20}{20}$, or \$240. A has 2 parts; B, 1 part; both, 3 parts. A has $\frac{2}{3}$ of \$240 = \$160; B has $\frac{1}{3}$ = \$80.

13. In $1\frac{2}{5}$ are $\frac{7}{5}$; $\frac{1}{5}=\frac{1}{7}$ of \$49 = \$7; $\frac{5}{5}$ = \$35. \$35 = $\frac{7}{100}$ of the principal; $\frac{1}{100}$ = \$5; $\frac{100}{100}$ = \$500. If twice A's money = 3 times B's, then once A's money = $1\frac{1}{2}$ times B's; B's, $\frac{2}{2}$; A's, $\frac{3}{2}$; both, $\frac{5}{2}$; $\frac{2}{2}$ = \$200, B's money; $\frac{3}{2}$ = \$300, A's money.

14. Two yr. 3 mo. = 27 mo. The int. for 1 mo. is $\frac{1}{27}$ of \$18 = \$$\frac{18}{27}$ = \$$\frac{2}{3}$; for 1 yr. 12 times \$$\frac{2}{3}$ = \$8. \$8 is 4 per cent, or $\frac{4}{100}$, of $\frac{3}{4}$ of A's and $\frac{1}{2}$ of B's; $\frac{1}{100}$ is \$2; $\frac{100}{100}$ is is \$200. $\frac{3}{4}$ of A's + $\frac{1}{2}$ of B's = \$200; but $\frac{1}{2}$ of A's = $\frac{2}{3}$ of B's, or $\frac{1}{4}$ of A's = $\frac{1}{3}$ of B's, and B's = $\frac{3}{4}$ of A's. Then, $\frac{6}{4}$ of B's = \$200, and B's money = \$133.33$\frac{1}{3}$. Since B's = $\frac{3}{4}$ of A's = \$133.33$\frac{1}{3}$, $\frac{1}{4}$ of A's = \$44.44$\frac{4}{9}$, and $\frac{4}{4}$ or A's money = \$177.77$\frac{7}{9}$.

LESSON LXXX.

1. One apple is worth $\frac{1}{8}$ of 24 plums = 3 plums; and 84 apples are worth 84 times 3 plums = 252 plums. One peach is worth $\frac{1}{12}$ of 252 plums = 21 plums; and 5 peaches are worth 105 plums.

2. Mary has 5 more than James, and Lucy 3 more than James. 5 + 3 = 8, and 32 — 8 = 24; $\frac{1}{3}$ of 24 = 8, James's share; 8 + 3 = 11, Lucy's; 8 + 5 = 13, Mary's.

3. Sixteen is twice the number; the number is 8.

4. C has $\frac{6}{6}$; B has $\frac{2}{6}$; A has $\frac{1}{6}$; C has $\frac{5}{6}$ more than A; $\frac{5}{6}$ = \$15; $\frac{1}{6}$ = \$3, A's; $\frac{6}{6}$ = \$18, C's; $\frac{2}{6}$ = \$6, B's.

5. Four fourths = James's money; $\frac{4}{4}+\frac{3}{4}=\frac{7}{4}$; \$34 — \$6 = \$28; \$28 = $\frac{7}{4}$; $\frac{1}{4}$ = \$4; $\frac{4}{4}$ = \$16, James's money; $\frac{3}{4}$ + \$6 = \$18, Thomas's money.

6. Eight eighths less $\frac{3}{8}=\frac{5}{8}$; $\frac{1}{9}$ of $\frac{5}{8}=\frac{5}{72}$; $\frac{4}{9}=\frac{20}{72}$; $\frac{5}{8}=\frac{45}{72}$; and $\frac{45}{72}+\frac{20}{72}=\frac{65}{72}$; $\frac{1}{72}=\frac{1}{65}$ of 65 sheep = 1 sheep; $\frac{72}{72}$ = 72 sheep.

7. One man will do the work in 12 da. of 10 hr., or in 120 da. of 1 hr. each; 8 men will do it in 15 da. of 1 hr., or in 2$\frac{1}{2}$ da. of 6 hr.

8. At 2 for 3 ct., 1 dozen cost 6 times 3 ct. = 18 ct.; at 2 for 5 ct., 1 doz. cost 6 times 5 ct. = 30 ct., and 2 doz. cost 18 ct. + 30 ct. = 48 ct. At 3 for 7 ct., 1 doz. sold for 4 times 7 ct. = 28 ct.; 2 doz. cost 56 ct.; and 56 ct. — 48 ct. = 8 ct., the gain on 2 doz.; 4 ct., gain on 1 doz.

9. Four horses, 2 mo. = 8 horses 1 mo.; 9 cows, 3 mo. = 27 cows, 1 mo.; 20 sheep, 5 mo. = 100 sheep, 1 mo. If 10 sheep = 2 horses, 5 sheep = 1 horse, and 100 sheep = 20 horses; 1 cow = $\frac{2}{3}$ of a horse, and 27 cows = 18 horses. Then A has the same as 8 horses; B, 18; and C, 20; and all have 46. A pays $\frac{8}{46}$ of \$92 = \$16; B, $\frac{18}{46}$ = \$36; C, $\frac{20}{46}$ = \$40.

10. He gave to each pair \$5; and \$5 in \$20 are contained 4 times. He had 4 sons and 4 daughters.

12. Nine less 3 = 6; 4 — 2 = 2; 6 ÷ 2 = 3, number of children.

14. One of John's steps = $1\frac{1}{4}$ of Henry's; 5 of John's = $6\frac{1}{4}$ of Henry's. He gains in taking 5 steps, $6\frac{1}{4}$ steps — 6 steps = $\frac{1}{4}$ step. He will take 4 times 5 steps = 20 steps to gain 1 step, and 7 times 20 steps = 140 steps to gain 7.

15. If 1 ox is worth 8 sheep, 3 oxen are worth 24 sheep, or 2 horses are worth 24 sheep; and 24 sheep are worth 24 times \$5 = \$120; 1 horse is worth $\frac{1}{2}$ of \$120 = \$60.

16. Two ct. + 24 ct. = 26 ct., and $\frac{1}{2}$ of 26 ct. = 13 ct., A's money; 13 ct. — 2 ct. = 11 ct., B's money.

17. Let $\frac{6}{6}$ = C's; $\frac{2}{6}$ = B's; $\frac{1}{6}$ = A's; then $\frac{6}{6} - \frac{2}{6} = \frac{4}{6}$, and $\frac{4}{6}$ = 20 yr. $\frac{1}{6}$ = 5 yr., A's age; $\frac{2}{6}$ = 10 yr., B's age; $\frac{6}{6}$ = 30 yr., C's age.

18. If \$15 is $\frac{3}{4}$ of their difference, then \$20 = the whole of the difference. If $\frac{2}{3}$ of A's = $\frac{4}{5}$ of B's, $\frac{3}{3}$ of A's = $\frac{6}{5}$ of B's; $\frac{3}{3} = \frac{5}{5}$, and $\frac{6}{5} - \frac{5}{5} = \frac{1}{5}$, their difference; and $\frac{1}{5}$ = \$20; $\frac{5}{5}$ = 100, B's; $\frac{6}{5}$ = \$120, A's.

19. One half of 17 is $8\frac{1}{2}$; and 10 less $8\frac{1}{2} = 1\frac{1}{2}$; and $1\frac{1}{2}$ in 15 is contained 10 times.

20. If 1 egg cost 2 ct., and 2 cost 6 ct., 3 cost 8 ct., and the average cost is $2\frac{2}{3}$ ct. 1 egg sells for $\frac{1}{3}$ of 10 ct. = $3\frac{1}{3}$ ct. The gain on 1 is $3\frac{1}{3} - 2\frac{2}{3} = \frac{2}{3}$; $\frac{2}{3}$ is $\frac{1}{4}$ of $2\frac{2}{3}$, or 25 per cent.

21. Eight less 5 = 3; 21 ÷ 3 = 7, number of playmates.

22. John gains 2 steps every time he takes 7; to gain 30 steps he must take 7 steps as many times as 2 is contained in 30, or 15 times; 15 times 7 steps = 105 steps.

23. Let $\frac{7}{7}$ = the watch, and $\frac{2}{7}$ = the chain; three times $\frac{2}{7}$ plus 2 times $\frac{7}{7} = \frac{20}{7}$, and $\frac{20}{7}$ = \$100; $\frac{1}{7}$ = \$5; $\frac{2}{7}$ = \$10, price of the chain; $\frac{7}{7}$ = \$35, price of the watch.

24. In $4\frac{1}{2}$ are $\frac{9}{2}$; $2\frac{4}{7} = \frac{18}{7}$; A does $\frac{2}{9}$ in 1 day; both do $\frac{7}{18}$ in 1 day; $\frac{7}{18}$ less $\frac{2}{9} = \frac{3}{18} = \frac{1}{6}$, what B does in 1 day. If B does $\frac{1}{6}$ in 1 day, he would do it all in 6 days.

25. He gave $\frac{1}{2}$ ct. each for the first lot, and $\frac{1}{4}$ ct. each for the second lot; for two he gave $\frac{1}{2}$ ct. + $\frac{1}{4}$ ct. = $\frac{3}{4}$ ct.; average price, $\frac{3}{8}$ ct. He sold them for $\frac{3}{5}$ ct. each; gain on each, $\frac{3}{5} - \frac{3}{8} = \frac{9}{40}$. If he gained $\frac{9}{40}$ ct. on one, to gain 18 ct. it took as many pears as $\frac{9}{40}$ ct. is contained times in 18 ct. = 80.

26. He receives \$59 — \$50 = \$9 for 2 years' interest on \$50, or \$4.50 for each year. $\frac{4.50}{50} = 9\%$.

27. She wished to buy as many yards as $\frac{1}{2}$ is contained times in $5 = 10$.

28. A's money $= \frac{5}{5}$; B's money $= \frac{2}{5} -$ \$5; $\frac{5}{5} + \frac{2}{5} -$ \$5 = \$51; $\frac{7}{5} =$ \$56; $\frac{1}{5} =$ \$8; $\frac{5}{5} =$ \$40, A's money; \$51 — \$40 = \$11, B's.

29. One third of the gain $= \frac{2}{15}$ of the selling price, and $\frac{3}{3} = \frac{6}{15}$, or $\frac{2}{5}$; $3\frac{3}{4}$ times \$4 = \$15, the cost. If the gain is $\frac{2}{5}$ of the selling price, then $\frac{5}{5} - \frac{2}{5} = \frac{3}{5}$, or the cost; $\frac{3}{5} =$ \$15; $\frac{5}{5} =$ \$25, the selling price.

30. The hound gains 5 of the hare's leaps every time the hare takes 3; to gain 100, he must take 3 leaps as many times as 5 is contained in $100 = 20$, and 20 times $3 = 60$.

31. Thomas's age = 3 parts; James's = 1 part; 3 parts — 1 part = 10, the difference. If 10 = 2 parts, then 5 yr. = James's age, and 15 yr. = Thomas's.

32. If $\frac{3}{7} = \frac{4}{5}$, then $\frac{7}{7} = \frac{28}{15}$; and $\frac{28}{15} + \frac{15}{15} = \frac{43}{15}$ of George's distance; $\frac{43}{15} = 86$ miles; $\frac{1}{15} = 2$ miles; $\frac{15}{15} =$ 30 miles, George's distance; $\frac{28}{15} = 56$ miles, John's distance.

34. The difference between selling the lot at 6 ct. a doz. and 10 ct. a doz. is 12 ct. + 18 ct. = 30 ct. The difference on 1 doz. is 10 ct. — 6 ct. = 4 ct. There were as many doz. as 4 ct. are contained times in 30 ct. $= 7\frac{1}{2}$ doz. The cost of the lot was $6 \times 7\frac{1}{2} + 12 = 57$ ct.; and the cost of 1 doz. was 57 ct. $\div 7\frac{1}{2} = 7\frac{3}{5}$ ct.

35. Let $\frac{10}{10}$ = A's age, and $\frac{5}{10}$ = B's; $\frac{3}{5}$ of $\frac{5}{10}$ are $\frac{3}{10}$, and $\frac{3}{10} + 44 = 2\frac{1}{2}$ times $\frac{10}{10} = \frac{25}{10}$. If $\frac{3}{10} + 44 = \frac{25}{10}$, then $\frac{22}{10} = 44$, and $\frac{1}{10} = 2$, and $\frac{10}{10} = 20$ yr., A's age; $\frac{5}{10} = 10$ yr., B's age.

36. Seven eighths of 24 miles = 21 miles. If 21 mi. are $\frac{3}{7}$, then $\frac{1}{7}$ is 7 mi., and $\frac{7}{7}$ are 49 mi., the distance from B to C; 49 mi. + 24 mi. = 73 mi., distance from A to C.

37. A, B, and C together can do $\frac{1}{4}$ in 1 da.; A and B together can do $\frac{1}{8}$ in 1 da.; B and C together can do $\frac{1}{6}$ in 1 da.; C can do in 1 da. $\frac{1}{4} - \frac{1}{8} = \frac{1}{8}$, and the whole in 8 da.; A can do in 1 da. $\frac{1}{4} - \frac{1}{6} = \frac{1}{12}$, and the whole in 12 da.; B can do in 1 da. $\frac{1}{6} - \frac{1}{8} = \frac{1}{24}$, and the whole in 24 da.

38. One duck cost $\$\frac{1}{6}$; 1 chicken, $\$\frac{1}{8}$, and 2 chickens, $\$\frac{2}{8}$; $\frac{1}{6} + \frac{2}{8} = \frac{10}{24}$; $\frac{1}{3}$ of $\frac{10}{24} = \frac{10}{72} = \frac{5}{36}$, the average cost. One third of $\frac{1}{2} = \frac{1}{6}$, the average selling price; $\frac{1}{6} = \frac{6}{36} - \frac{5}{36} = \frac{1}{36}$, the average gain; the whole gain was $\$2\frac{1}{2} = \$\frac{5}{2}$; $\frac{5}{2} = \frac{90}{36}$; $\frac{90}{36} \div \frac{1}{36} = 90$, the whole number; $\frac{2}{3}$ of $90 = 60$, the chickens; $\frac{1}{3}$ of $90 = 30$, the ducks.

39. Eight ct. − 3 ct. = 5 ct.; 6 ct. + 29 ct. = 35 ct.; 35 ct. = $\frac{5}{8}$ of cost of oranges; $\frac{1}{8}$ = 7 ct., and $\frac{8}{8}$ = 56 ct.; 56 ct. − 6 ct. = 50 ct., James's money.

40. A rides $\frac{1}{5}$ of 10 miles in $\frac{1}{4}$ of an hour, and 8 miles in 1 hour; A will travel 18 miles in $18 \div 8 = 2\frac{1}{4}$ hr. B travels $\frac{1}{8}$ of a mile in $\frac{1}{5}$ hr., and 5 miles an hour; B will travel $2\frac{1}{4}$ times 5 mi. = $11\frac{1}{4}$ mi., while A travels 18 mi.

41. Three halves + $\$2\frac{1}{2} = \$40$; then $\frac{3}{2} = \$37\frac{1}{2}$; $\frac{1}{2} = \$12\frac{1}{2}$; $\frac{2}{2} = \$25$, his money.

42. C received $\frac{21}{21}-\frac{6}{21}-\frac{7}{21}=\frac{8}{21}$; $\frac{8}{21}-\frac{6}{21}=\frac{2}{21}$; $\frac{2}{21}=\$160$; $\frac{1}{21}=\$80$; $\frac{6}{21}=\$480$, A's legacy; $\frac{7}{21}=\$560$, B's legacy; $\frac{8}{21}=\$640$, C's legacy.

43. Both consume $\frac{6}{15}$ in 6 days, and $\frac{15}{15}$ less $\frac{6}{15}=\frac{9}{15}=\frac{3}{5}$ remaining. The woman consumes $\frac{1}{24}$ of $\frac{3}{5}$ in one day $=\frac{3}{120}=\frac{1}{40}$, and all in 40 days. Both consume $\frac{1}{15}$ in one day; $\frac{1}{15}-\frac{1}{40}=\frac{5}{120}=\frac{1}{24}$, what the man consumes in one day. It would last him alone 24 days.

44. Three and one half ct. $+\ 6\frac{1}{2}$ ct. $=10$ ct., the price of 2 pounds of the mixture; $100\div10=10$; 10 times 2 pounds $=20$, the number of pounds.

45. Let $\frac{1}{10}=$ C's age; $\frac{2}{10}=$ B's; and $\frac{10}{10}=$ C's. $\frac{10}{10}-\frac{1}{10}=\frac{9}{10}$; $\frac{9}{10}=45$ yr.; $\frac{1}{10}=\frac{1}{9}$ of 45 yr. $=5$ yr., C's age; $\frac{2}{10}=10$ yr., B's age; $\frac{10}{10}=50$ yr. $=$ A's age.

46. Three fifths $=$ Mary's age, and $\frac{5}{5}=$ Ella's; their sum is $\frac{8}{5}$; twice Ella's is $\frac{10}{5}$; $\frac{10}{5}-\frac{8}{5}=\frac{2}{5}$, and $\frac{2}{5}=6$ yr.; $\frac{1}{5}=3$ yr.; $\frac{5}{5}=15$ yr., Ella's age; $\frac{3}{5}=9$ yr., Mary's age.

47. Both do $\frac{4}{16}$ in 4 days; and $\frac{16}{16}-\frac{4}{16}=\frac{3}{4}$, that B finishes in 36 days. In one day he does $\frac{1}{36}$ of $\frac{3}{4}=\frac{3}{144}=\frac{1}{48}$, and he does all in 48 days; $\frac{1}{16}-\frac{1}{48}=\frac{1}{24}$, what A does in 1 day, and he would do all in 24 days.

48. Three doz. at 1 ct. each $=36$ ct.; 2 doz. at 4 eggs for 3 ct. $=18$ ct.; 2 doz., the remainder, at 4 eggs for 5 ct. $=30$ ct.; 36 ct. $+$ 18 ct. $+$ 30 ct. $=84$ ct.; $84=7$ doz.; $\frac{1}{7}$ of 84 ct. $=12$ ct. a doz.

50. If he had worked 30 days, he would have received 30 times 30 ct. = \$9. Each day he is idle he gives 20 ct. for board and forfeits 30 ct. for not working = 50 ct. \$9 — \$5 = \$4; \$4.00 ÷ 50 ct. = 8, number of days idle; 30 days — 8 days = 22, number of days he worked.

51. The difference per yard is $2\frac{1}{2}$ ct.; $40 \div 2\frac{1}{2} = 16$, the number of yards.

52. If 4 of Moses's steps = 7 of Noah's, then one of Moses's = $1\frac{3}{4}$ of Noah's, and 5 of Moses's = $8\frac{3}{4}$ of Noah's; Moses gains $8\frac{3}{4} - 7 = 1\frac{3}{4}$ of Noah's steps every time he takes 5 steps; to gain 35 he must take 5 steps as many times as $1\frac{3}{4}$ are contained in 35. $35 = \frac{140}{4}$; $1\frac{3}{4} = \frac{7}{4}$; $\frac{140}{4} \div \frac{7}{4} = 20$; 20 times 5 steps = 100 steps.

53. One man will do as much work as 6 boys; 2 men as much as 12 boys; then 2 men would do in a week as much as 12 boys, and to do it in 1 day it would take 6 times 2 men = 12 men.

54. Let $\frac{1}{12}$ = the number in the first field; $\frac{4}{12}$, in the second; and $\frac{12}{12}$, in the third; then $\frac{12}{12} - \frac{1}{12} - \frac{4}{12} = \frac{7}{12}$, and $\frac{7}{12} = 70$; $\frac{1}{12} = 10$, the number in the first field; $\frac{4}{12} = 40$, the number in the second; $\frac{12}{12} = 120$, the number in the third.

55. For 24 days he would have received \$48. He loses \$2 each day he is idle, and pays 50 ct. for board = $\$2\frac{1}{2}$; \$48 — \$38 = \$10; $\$10 \div \$2\frac{1}{2} = 4$, number of days idle; 24 — 4 = 20, number of days he worked.

56. Since \$12 = $\frac{2}{7}$ of B's and C's, $\frac{7}{7}$ = \$42; and \$42 + \$12 = \$54, what all had. If $\frac{3}{8}$ of C's = $\frac{3}{10}$ of A's and B's,

then $\frac{1}{8}=\frac{1}{10}$, and the whole of C's $=\frac{8}{10}$ of A's and B's; hence, $\frac{18}{10}$ of A's and B's $=$ \$54; $\frac{1}{10}=$ \$3, and $\frac{10}{10}=$ \$30; \$30 — \$12 $=$ \$18, B's share; \$54 — \$30 $=$ \$24, C's share.

57. Six and six sevenths pounds cost $6\frac{6}{7}$ times 8 ct. $=$ $54\frac{6}{7}$ ct.; $\frac{1}{6}$ of $54\frac{6}{7}$ ct. $=9\frac{1}{7}$ ct.; $54\frac{6}{7}$ ct. $+ 9\frac{1}{7}$ ct. $=64$ ct.

58. The first has \$1200 for 1 mo.; the second has \$2400 for 1 mo., or \$1200 more than the first; the first must put in for the remaining 6 mo. $\frac{1}{6}$ of \$1200 $=$ \$200.

59. The difference between selling at 9 ct. and 12 ct. is \$1.50 on the whole; on one pound the difference is 3 ct.; there are as many pounds as \$1.50 $\div$ 3 $=$ 50.

60. One of B's steps $=1\frac{1}{2}$ of A's, and 4 of B's $=6$ of A's; B gains $6-5=1$ step every time he takes 4 steps; B takes 4 steps 9 times in taking 36 steps. If he gains 1 step every time he takes 4, then A is 9 steps in advance of B.

61. Both had 9 oranges; each ate $\frac{1}{3}$ of $9=3$. Thomas ate 2 of John's oranges, and should give him $\frac{2}{3}$ of 9 ct. $=$ 6 ct.; he ate one of James's, and should give him $\frac{1}{3}$ of 9 ct. $=3$ ct.

SOLUTIONS

TO

QUESTIONS AND PROBLEMS

IN

RAY'S NEW PRACTICAL ARITHMETIC.

NOTATION.

NUMBERS TO BE WRITTEN.

Art. 5.

(1.) 23; 24; 25; 26; 27; 28; 29.
(2.) 37; 42; 56; 69; 73; 87; 94.
(3.) 83; 45; 99; 51; 36; 78; 62.
(4.) 55; 93; 81; 67; 49; 74; 38.
(5.) 76; 44; 82; 57; 35; 91; 63.

NUMBERS TO BE READ.

(1.) Seventy-one; thirty-two; fifty-three; eighty-four; sixty-five; forty-six; ninety-seven.

(2.) Fifty-eight; thirty-four; seventy-nine; sixty-six; forty-one; eighty-five; ninety-two.

(3.) Seventy-five; forty-three; eighty-eight; sixty-one; fifty-nine; thirty-three; ninety-five.

(4.) Thirty-nine; seventy-two; fifty-four; eighty-six; forty-seven; ninety-eight; sixty-four.

(5.) Sixty-eight; seventy-seven; thirty-one; eighty-nine; fifty-two; ninety-six; forty-eight.

NUMBERS TO BE WRITTEN.

Art. 7.

(1.) 130; 140; 150; 160; 170; 180.

(2.) 123; 456; 789; 147; 258; 369.

(3.) 102; 345; 678; 234; 567; 890.

(4.) 453; 786; 912; 230; 450; 670.

(5.) 153; 486; 729; 103; 406; 709.

NUMBERS TO BE READ.

(1.) Two hundred and ten; three hundred and twenty; four hundred and thirty, etc.

(2.) Two hundred and thirteen; five hundred and forty-six; eight hundred and seventy-nine; four hundred and seventeen, etc.

(3.) Two hundred and one; four hundred and thirty-five; seven hundred and sixty-eight; three hundred and twenty-four, etc.

Art. 11.

(2.) 2000; 30000; 400000.

(3.) 5000000; 60000000; 700000000.

(4.) 8000000000; 90000000000; 100000000000.

(5.) 1200; 2100.

(6.) 3450; 6789.

(7.) 12345.

(8.) 678912.

(9.) 1357924.

(10.) 68143792.

(11.) 1001; 1010; 1100.

(12.) 1101; 1110; 1111.

(13.) 2003; 4050.

(14.) 45026.

(15.) 80201.

(16.) 90001.
(17.) 410205.
(18.) 100010.
(19.) 3070509.
(20.) 45083026.
(21.) 909090000.
(22.) 700010002.
(23.) 40000200005.
(24.) 726050001243.
(25.) 80703000504.

NUMERATION.

Art. 12.

(2.) Forty-one thousand five hundred and eighty-two; seven hundred and sixty-three thousand four hundred and ninety-one; two million five hundred and nineteen thousand eight hundred and thirty-four; three hundred and seventy-five million four hundred and eighty-six thousand nine hundred and twenty-one; four billion nine hundred and twenty-three million one hundred and seventy-six thousand three hundred and fifty-eight.

(3.) Thirty-seven billion five hundred and eighty-four million two hundred and sixteen thousand nine hundred and seventy-four; four hundred and thirty-two billion six hundred and eighty-five million seven hundred and twenty-nine thousand one hundred and forty-five; six trillion two hundred and fifty-three billion nine hundred and seventy-one million four hundred and thirty-eight thousand two hundred and sixty-seven.

(4.) One thousand three hundred; two thousand five hundred and forty; six thousand and seventy; eight thousand and nine; thirteen thousand two hundred; one thousand and five.

(5.) Six hundred and eighty-two thousand three hundred; eight million six hundred thousand and fifty; three thousand and forty; fifty thousand and four; seven hundred and four thousand two hundred and eight.

(6.) Seven thousand and eighty-five; sixty-two thousand and one; four hundred thousand and nine; two million one hundred and two thousand one hundred and two; nine million one thousand and three.

(7.) One hundred and thirty million six hundred and seventy thousand nine hundred and twenty-one; six billion nine hundred million seven hundred and two thousand and three; twenty-three billion four million ninety thousand seven hundred and one; nine billion four hundred and twenty million one hundred and sixty-three thousand and seventy.

(8.) Five hundred and seventy trillion ten million three hundred and twenty-six thousand and forty-nine; two hundred quadrillion one hundred and three trillion four hundred and seventy-eight billion five hundred and eleven million nine hundred and ninety-two thousand four hundred and eighty-five.

(9.) Forty-five quintillion seven hundred and sixty-three quadrillion twenty billion one hundred and eight million five hundred and seven.

(10.) Eight hundred trillion eight hundred and twenty billion twenty million eight hundred and two thousand and eight.

Art. 13. (P. 21.)

(1.) I, II, III, IV, V, VI, VII, VIII, IX, X, XI, XII, XIII, XIV, XV, XVI, XVII, XVIII, XIX, XX.

(2.) XXI, XXII, XXIII, XXIV, etc., XXX.

(3.) XXX, XL, L, LX, LXX, LXXX, XC.

(4.) LVII, XXIX, LXI, XXXVIII, XLVI, LXXII, XCIII.

(5.) C, CI, CVI, CXVII, CXXIX, CLXVIII.

(6.) CXCIX, CCXLVI, CCCIX, CCCCLXXXII, DXXVII, DCXCIII.

(7.) DCCXXXIV, DCCCLIX, DCCCCLXXV, MI, MX.

(8.) MXLVIII, MCXIX, MCCLXXXV, MCCCXXVI.
(9.) MCCCCXCII, MDCCLXXVI, MDCCCLXI, MDCCCC.

ADDITION.

Art. 17.

(2.)	(4.)
$210	50230
142	3105
35	423
$387	53758

Art. 19.

(8.) 21023. (9.) 27910. (10.) 89569.

(11.) 2499593. (12.) 24194086.

(29.) $146 + 607 + 47 = 800$: 1700 yr. + 800 yr. = 2500 yr.

(30)	(31)	(32)	(33)
3005	275432	880000889	8955752
42627	402030	2002002	6917246
105	300005	77436000	94523
307004	872026	206005207	15967521
80079	4002347	49003	
320600	5851840	990019919	
753420		2155513020	

(34)	(35)	(36)	(37)	(38)
$600	$7850	$8785	$7000	30
1325	3275	12789	12875	30
30	3275	878	5600	25
120	2650	1250	4785	25
250	2650	$23702	3500	25
140	2650		$33760	25
120	$22350			20
115				20
$2700				200 yd.

SUBTRACTION.

Art. 26.

(5)	(6)	(8)	(9)
4444444	91516171	153425178	100000000
1234567	15161718	53845248	10001001
3209877	76354453	99579930	89998999

(14)	(15)	16)	(18)	(19)
$1840	$10104	$100000	912010	4000000
475	7426	11	50082	4004
$1365	$2678	$99989	861928	3995996

(20)	(21)	(22)	(23)
2020930	2000687	17102102	$30000
1009006	405022	13000201	26967
1011924	1595665	4101901	$3033

(24)	(25)
18126402	19900900900
9238715	9909090009
8887687	9991810891

Addition and Subtraction.

(1)	(2)	(3)		(4)
275	6723	$2675	$3000	$250
381	479	4375	4947	650
625	6244	1897	$7947	$900
1281	347	$8947		
1098	5897	7947		$1300
183	228	$1000, *Ans.*		900
	5669			$400

(5)

$450	
725	
1235	
4675	$5935
1727	877
$8812	$6812
6812	
$2000, *Ans.*	

(6)

$350
125
375
150
$1000
$2300
1000
$1300, *Ans.*

(7)

$4875
4875 }
2250 }
3725
$15725
$20838
15725
$5113, *Ans.*

(8)

$16785	$49570
24937	41722
$41722	$7848, *Ans.*

(9)

$7895	$10093
175	8073
3	$2020, *Ans.*
$8073	

(10)

$5750	$10000
925	8925
1575	$1075, *Ans.*
675	
$8925	

(11)

$4625	$6955	$9395
3785	895	9225
985	1375	$170, *Ans.*
$9395	$9225	

(12)

Received, $50	Spent, $25	$100
50	7	91
$100	2	$9, *Ans.*
	5	
	35	
	7	
	2	
	8	
	$91	

MULTIPLICATION.

Art. 31.

(25)	(26)	(27)	(28)	(29)
235	346	425	518	279
13	19	29	34	37
705	3114	3825	2072	1953
235	346	850	1554	837
3055	6574	12325	17612	10323

(30)	(31)	(32)	(33)	(34)
869	294	429	485	624
49	57	62	76	85
7821	2058	858	2910	3120
3476	1470	2574	3395	4992
42581	16758	26598	36860	53040

(35)	(36)	(37)	(38)
976	342	376	476
97	364	526	536
6832	1368	2256	2856
8784	2052	752	1428
94672	1026	1880	2380
	124488	197776	255136

(39)	(40)	(41)	(42)
2187	3489	1646	8432
215	276	365	635
10935	20934	8230	42160
2187	24423	9876	25296
4374	6978	4938	50592
470205	962964	600790	5354320

(43)
6874
829
61866
13748
54992
5698546

(44)
2873
1823
8619
5746
22984
2873
5237479

(45)
4786
3497
33502
43074
19144
14358
16736642

(46)
87603
9865
438015
525618
700824
788427
864203595

(47)
83457
6835
417285
250371
667656
500742
570428595

(48)
31624
7138
252992
94872
31624
221368
225732112

(51)
675
13
2025
675
8775 ct.

(52)
496
24
1984
992
11904 ct.

(53)
152
28
1216
304
4256 mi.

(54)
1760
209
15840
3520
367840 yd.

(55)
365
24
1460
730
8760
8
70080 mi.

(56)
2029
1007
14203
2029
2043203

(57)
80401
60007
562807
482406
4824622807

(58)
101032
20001
101032
202064
2020741032

(60)
36
45
180
144
1620

36
55
180
180
1980
1620
Ans. 360 ct.

(61)

95 ct. — 2 ct. = 93 ct.

```
  2650
    93
  7950
 23850
 246450 ct.
```

(62)

$75 × 6 = $450
125 × 5 = 625
$1075

$150 × 11 = $1650
$1650 — $1075 = $575, *Ans.*

(63)

```
              $250
$325 × 2 =     650
 175 × 3 =     525
             $1425
               356
             $1781, Ans.
```

(64)

```
24 × $5 =  $120
36 × 14 =   504
 9 × 45 =   405
          $1029
            275
           $754, Ans.
```

(65)

```
  75     85
  37     54
 525    340
225    425
2775   4590
4590
7365
5284
2081, Ans.
```

(66)

```
  69     48
  53     27
 207    336
345     96
3657   1296
       3657
       4953
       4279
        674, Ans.
```

(67)

```
   63 lb.
   50
 3150 lb.

  3150
    34 ct.
 12600
 9450
107100 ct., Ans.
```

Art. 32.

(2)			(3)	(4)
$124		$124	1512 mi.	2873 lb.
6		8	8	9
$744	or	$992	12096 mi.	25857 lb.
4		3	7	6
$2976		$2976	84672 mi.	155142 lb.

(5)	(6)
2874	8074
9	12
25866	96888
8	9
206928	871992

Art. 33.

(1)	(2)	(3)
245	138	428
100	1000	10000
24500	138000	4280000

(4)	(5)	(6)
872	9642	10045
100000	1000000	1000000
87200000	9642000000	10045000000

Art. 34.

(3)	(4)	(5)	(6)
2350	80300	10240	9600
60	450	3200	2400
141000	4015	2048	384
	3212	3072	192
	36135000	32768000	23040000

(7)	(8)	(9)
18001	8602	3007
26000	1030	9100
108006	25806	3007
36002	8602	27063
468026000	8860060	27363700

(10)	(11)	(12)
80600	70302	904000
7002	80300	10200
1612	210906	1808
5642	562416	904
564361200	5645250600	9220800000

SHORT DIVISION.

Art. 41.

(24)	(25)	(29)
3) 894	4) 140	4) 321276
298	35	80319

(32)	(33)	(46)
11) 495	9) 3582	4) 144
45	398	3) 36
		12 *Ans.*

(47)		(48)		
5) 195	3) 39	8) 192	11) 275	25
39	13, *Ans.*	24	25	24
				1, *Ans.*

LONG DIVISION.

Art. 42.

(5)

14) 11577 ($826\frac{13}{14}$, *Ans.*
112
37
28
97
84
13

(6)

15) 48690 (3246
45
36
30
69
60
90
90

(7)

23) 1110960 ($48302\frac{14}{23}$
92
190
184
69
69
60
46
14

(8)

67) 122878 (1834
67
558
536
227
201
268
268

(9)

53) 12412 ($234\frac{10}{53}$
106
181
159
222
212
10

(10)

72) 146304 (2032
144
230
216
144
144

(11)

54) 47100 ($872\frac{12}{54}$
432
390
378
120
108
12

(12)

88) 71104 (808
704
704
704

(13)

66) 43956 (666
396
435
396
396
396

(14)

99) 121900 ($1231\frac{31}{99}$
99
229
198
310
297
130
99
31

(15)

112) 25312 (226
224
291
224
672
672

(16)

123) 381600 ($3102\frac{54}{123}$
369
126
123
300
246
54

(17)

204) 105672 (518
1020
367
204
1632
1632

(18)

1234) 600000 ($486\frac{276}{1234}$
4936
10640
9872
7680
7404
276

(19)

4321) 1234567 ($285\frac{3082}{4321}$
8642
37036
34568
24687
21605
3082

(20)

819) 50964242 (6518
46914
40502
39095
14074
7819
62552
62552

(21)

9876) 48905952 (4952
39504
94019
88884
51355
49380
19752
19752

(22)

12345) 4049160 (328
37035
34566
24690
98760
98760

(23)

```
973) 552160000 (567482 14/973
     4865
     ----
      6566
      5838
      ----
       7280
       6811
       ----
        4690
        3892
        ----
         7980
         7784
         ----
          1960
          1946
          ----
            14
```

(24)

```
15) 3465 (231
    30
    --
     46
     45
     --
      15
      15
      --
```

(25)

```
26) 364 (14 days.
    26
    ---
    104
    104
    ---
```

(26)

```
19) 1083 (57 dollars.
    95
    ---
    133
    133
    ---
```

(27)

```
107) 9523 (89 bu.
     856
     ---
      963
      963
      ---
```

(28)

```
63) 14868 (236 hhd.
    126
    ---
     226
     189
     ---
      378
      378
      ---
```

(29)

```
365) 50000 (136
     365
     ----
     1350
     1095
     ----
      2550
      2190
      ----
       360
```

Ans. $136 and $360 over.

(30)

```
365) 379600 ($1040
     365
     ----
      1460
      1460
      ----
         0
```

(31)

```
1235) 6571435 (5321
      6175
      ----
       3964
       3705
       ----
        2593
        2470
        ----
         1235
         1235
         ----
```

(32)

```
405) 1247400 (3080
     1215
     ----
       3240
       3240
       ----
          0
```

(33)

1006) 10401000 ($10338\frac{972}{1006}$
1006
3410
3018
3920
3018
9020
8048
972

(34)

684) 109440 (160 A.
684
4104
4104
0

(35)

56) 8288 (148 A.
56
268
224
448
448

(36)

269) 262275 (975 dollars.
2421
2017
1883
1345
1345

(37)

24) 24899 ($1037\frac{11}{24}$ mi.
24
89
72
179
168
11

(38)

238) 3731840 (15680 dollars.
238
1351
1190
1618
1428
1904
1904
0

(39)

24) 27048 (1127 ft.
24
30
24
64
48
168
168

(40)

11520000) 92160000 (8 min.
92160000

(41)

94231
86247
16) 7984 (499, *Ans.*
64
158
144
144
144

(42)
46712
6848
104) 53560 (515
520
156
104
520
520

(43)
497
583
1491
3976
2485
71) 289751 (4081
284
575
568
71
71

(44)
2832
987
1845
678
87) 2523 (29
174
783
783

(45)
4896
2384
2512
49
22608
10048
112) 123088 (1099
112
1108
1008
1008
1008

(46)
228
786
1014
95
5070
9126
114) 96330 (845
912
513
456
570
570

(47)
478 478
296 296
182 774
182
1548
6192
774
387) 140868 (364
1161
2476
2322
1548
1548

(48)
7560
3885
175) 3675 (21 horses.
350
175
175

(49)
7350
4655
49) 12005 (245 A.
98
220
196
245
245

(50)
58
77
406
406
4466

5742
4466
58) 1276 (22 dollars
116
116
116

(51)
240
26
1440
480
6240
2820
180) 3420 (19 horses.
180
1620
1620

(52)
125 lots.
250 dolls. each.
6250
250
31250
10625
$20625, whole gain.

25) 20625 (825 dolls., gain per acre.
200
62
50
125
125

125) 20625 (165 dolls., gain on each lot.
125
812
750
625
625

Art. 43.

(3)
9) 2583
7) 287
Ans. 41

(4)
4) 6976
8) 1744
Ans. 218

(5)
4) 2744
7) 686
Ans. 98

(6)
6) 6145
7) 1024—1 rem.
146—2
$6 \times 2 + 1 = 13$ rem.
Ans. $146\frac{13}{42}$

(7)

11) 19008

12) 1728

Ans. 144

(8)

8) 7840

8) 980

122—4

$8 \times 4 = 32$ rem.

Ans. $122\frac{32}{64}$

(9)

8) 14771

9) 1846—3

205—1

$8 \times 1 + 3 = 11$ rem.

Ans. $205\frac{11}{72}$

(10)

9) 10206

9) 1134

Ans. 126

(11)

11)81344

11)7394—10

672— 2

$2 \times 11 + 10 = 32$

Ans. $672\frac{32}{121}$

or

121) 81344 ($672\frac{32}{121}$

726

874

847

274

242

32

(12)

9) 98272

12) 10919—1

909—11

$11 \times 9 + 1 = 100$

Ans. $909\frac{100}{108}$

or

108) 98272 ($909\frac{100}{108}$, *Ans.*

972

1072

972

100

Art. 44.

(2)

1|0)268|2

$268\frac{2}{10}$

(3)

1|00)47|00

47

(4)

1|00)372|01

$372\frac{1}{100}$

(5)

1|00)462|50

$462\frac{50}{100}$

(6)

1|000)18|003

$18\frac{3}{1000}$

Art. 45.

(4)

4|000)73|005

$18\frac{1005}{4000}$

(5)

9|000)36|001

$4\frac{1}{9000}$

(6)

11|000)1078|000

98

(7)

```
18|0)4016|7(223 27/180
     36
     --
      41
      36
      --
       56
       54
       --
        2
```

(8)

```
21|00)9072|37(432 37/2100
      84
      --
       67
       63
       --
        42
        42
        --
```

(9)

```
64|00)3640|06(56 5606/6400
      320
      ---
       440
       384
       ---
        56
```

(10)

```
25|0000)7654|6037(306 46037/250000
        75
        --
         154
         150
         ---
           4
```

(11)

```
634|00)435637|54(687 7954/63400
       3804
       ----
        5523
        5072
        ----
         4517
         4438
         ----
           79
```

Art. 49.

(1)

```
 $96   $500
 120    271
  55   ----
----   $229, Ans.
$271
```

(2)

```
$243 1st.   $243    $1265
  61         304      772
----         225    -----
$304 2d.    ----     $493, Ans.
  79        $772
----
$225 3d.
```

(3)

```
157     428
264     186
305     ---
 97     614
123
---
946
614
---
332, Ans.
```

(4)

```
 9503    57068
  586    16967
 4794    -----
 1234    40101, Ans.
  850
-----
16967
```

(5)

```
$12307
  8706
------
$21013, am't with gain.

 $237
  301
 5380
-----
$5918, am't spent.

$21013
  5918
------
$15095, Ans.
```

(6)

```
86)31173(362 41/86
   258
   ---
    537
    516
    ---
     213
     172
     ---
      41
```

(7)

```
 28
  3   $
---
25)1400(56 dolls.
   125
   ---
    150
    150
    ---
```

(8)

```
   63 gallons.
    5
  ---
15)315(21, Ans.
   30
   --
    15
    15
    --
```

(9)

```
    73900
       70
    -----
214)73830(345, Ans.
    642
    ---
     963
     856
     ---
     1070
     1070
     ----
```

(10)

```
148       148
 56        56
---       ---
 92       204, sum.
           92, diff.
          ---
          408
         1836
         ----
     23)18768(816, Ans.
        184
        ---
          36
          23
          --
          138
          138
          ---
```

(11)

```
 $60        $45
   8         14
----        ---
$480        180
 630        45
----        ----
6)1110      $630
 ----
  185       Ans., 185 yards.
```

(12)

$30
70
$2100

$6000
2100
$3900

3900 ÷ 25 = 156

156 acres.
70 acres.
226 acres, *Ans.*

(13)

$360
300
150
100
90
$1000, spent each yr.

$1800
1000
$800, saved each yr.

yr.
800) 10400 (13, *Ans.*
800
2400
2400

(14)

40 × $15 = $600
80 × 25 = 2000
Amt. pd., $2600

120 — 90 = 30 acres.
30 × 60 = 1800
$4500 + $1800 = $6300, 1*st Ans.*

$6300
2600
$3700, gain.

(15)

275 × $4 = $1100

250 × $5 = $1250
25 × 6 = 150
$1400
1100
$300, *Ans.*

(16)

125
$85
625
1000
$10625

75
$115
375
75
75
$8625
10625
$19250

175) 19250 (110, *Ans.*
175
175
175
0

(17)

$150
15
750
150
$2250
2500
$4750
2250
$2500
95
$2405, *Ans.*

$125
20
$2500

$45
50
$2250

COMPOUND NUMBERS.

U. S. Money.

EXAMPLES TO BE WRITTEN.

Art. 53.

(1.) $12.178
(2.) $ 6.066
(3.) $ 7.007
(4.) $40.535
(5.) $ 2.03
(6.) $ 20.022
(7.) $100.10
(8.) $200.02
(9.) $400.018

EXAMPLES TO BE READ.

Eighteen dollars sixty-two cents five mills; twenty dollars thirty-two cents four mills; seventy-nine dollars five cents; forty-six dollars; seventy dollars one cent five mills; one hundred dollars twenty-eight cents; etc.

Art. 55.

REDUCTION.

Since the operations in this section consist simply in adding ciphers or removing them, or erasing points or inserting them between the different denominations, it is deemed unnecessary to occupy space, as the whole solution, when presented to the eye, would consist in nothing more than writing down the question to be solved, and then placing the answer under it.

Art. 56.

ADDITION.

(2)	(3)	(4)
$17.15	$18.041	$43.75
23.43	16.317	29.18
7.19	100.503	17.63
8.37	87.338	268.95
12.31	$222.199	718.07
$68.45		$1077.58

(5)
$200.00
43.87
56.93
8.50
2.31
―――
$311.61

(6)
$504.06
420.19
105.50
304.00
888.47
―――
$2222.22

(7)
$5.070
30.203
100.005
60.020
700.011
1000.100
40.004
64.587
―――
$2000.000

SUBTRACTION.

Art. 57.

(2)
$29.342
17.265
―――
$12.077

(3)
$46.28
17.75
―――
$28.53

(4)
$20.05
5.50
―――
$14.55

(5)
$3.00
.03
―――
$2.97

(6)
$10.000
.001
―――
$9.999

(7)
$50.000
.505
―――
$49.495

(8)
$1000.000
1.011
―――
$998.989

(9)
$1000.43
900.68
―――
$99.75

MULTIPLICATION.

Art. 58.

(2)
$7.835
8
―――
$62.680

(3)
$12.093
9
―――
$108.837

(4)
$23.018
16
―――
138108
23018
―――
$368.288

(5)
$35.14
53
―――
10542
17570
―――
$1862.42

(6)
$125.02
62
―――
25004
75012
―――
$7751.24

(7)
$40.04
102
―――
8008
4004
―――
$4084.08

```
   (8)          (9)          (10)          (11)
 $0.125       $3.28         $1.06          $5.75
     17          38           338             38
 ------       -------      -------        -------
    875        2624           848           4600
   125         984           318           1725
 ------       -------        318          -------
 $2.125       $124.64      -------        $218.50
                           $358.28
```

```
   (13)         (14)         (15)          (16)
  $0.34        $5.67        $2.69          $1.25
     89           24          169            691
 ------       -------      -------        -------
    306         2268         2421            125
   272         1134         1614            1125
 ------       -------       269             750
 $30.26       $136.08      -------        -------
                           $454.61        $863.75
```

```
        (17)                 (18)
          73                  281 lb.
          63 gal.               4
        ----               --------
         219                 1124 lb.
        438                $0.065
        ----              ---------
        4599 gal.            5620
       $0.55                6744
     --------             ---------
       22995              $73.060
      22995
     --------
     $2529.45
```

```
  (19)           (20)                          (21)
   35         312        3432           18          $1.25
   10 yd.      11 hr.   $0.13            3 bu.         54
 ----        ----      ------          ----         ------
  350 yd.    3432 hr.   10296           54 bu.        500
 $0.01                  3432                          625
 -----                 ------                       ------
 $3.50                 $446.16, Ans.           Ans. $67.50
```

```
     (22)
  $10.001
      150
  -------
   500050
  10001
 --------
$1500.150
```

```
        (23)
   17        $0.247
   51 lb.       867
 ----        ------
   17          1729
  85          1482
 ----        1976
  867 lb.   --------
            $214.149, Ans.
```

DIVISION.

Case I.

Art. 59.

```
(2)
9)72
  --
  8 lb.
```

```
(3)
375)6000(16
    375
    ----
    2250
    2250
    ----
```

```
(4)
8)280
  ----
  35 yd.
```

```
(5)
25)300(12 yd.
   25
   --
    50
    50
    --
```

```
(6)
805)16100(20 bbl.
    1610
    ----
       0
```

```
(7)
75)1200(16
   75
   ---
   450
   450
   ---
```

```
(8)
1125)234000(208 bu.
     2250
     ----
      9000
      9000
      ----
```

Case II.

```
(3)
8)$65.000
  -------
   $8.125
```

```
(4)
23)$29.610($1.287+
    23
    --
     66
     46
     ---
     201
     184
     ---
      170
      161
      ---
        9
```

```
(5)
4)$92.250
  --------
  $23.062+
```

```
(6)
8)$57.500
  -------
  $7.187+
```

(7)

```
16)$25.76($1.61
   16
   --
    97
    96
    --
     16
     16
     --
```

(8)

```
755)$328.425($0.435
     3020
     ----
      2642
      2265
      ----
       3775
       3775
       ----
```

(9)

```
313)$800.000($2.555+
    626
    ---
    1740
    1565
    ----
     1750
     1565
     ----
      1850
      1565
      ----
       285
```

(10)

```
133)$10000.000($75.187+
     931
     ---
      690
      665
      ---
       250
       133
       ---
       1170
       1064
       ----
        1060
         931
        ----
         129
```

(11)

```
154)$2705.010($17.565
     154
     ---
     1165
     1078
     ----
       870
       770
       ---
      1001
       924
       ----
        770
        770
        ---
```

(12)

```
      25
      15 lb.
     ---
     125
     25
     ---
lb. 375)$60.00($0.16
         375
         ---
         2250
         2250
         ----
```

(13)

```
    235 lb.
      8
    ----
lb. 1880)$122.200($0.065
          11280
          -----
           9400
           9400
           ----
```

Art. 60.

(1)
$47.50
38.45
15.47
19.43
$120.85

(2)
$35.25
23.75
$59.00
59.00
1.00
$119.00

(3)
$18.38
81.62
$100.00
200.00
$300.00

(4)
$0.75
.35
.50
1.50
$3.10

$5.00
3.10
$1.90, *Ans.*

(5)
$8.10
5.65
$0.25 × 8 = 2.00
4.00
$19.75

$20.00
19.75
$0.25, *Ans.*

(6)
$50.00
30.50
$19.50
6
$117.00

(7)
$3.85
1.25
2.50
1.50
$9.10

$21.75
9.10
$12.65, *Ans.*

(8)
$37.06
200.85
400.00
236.75
124.34
$999.00
889.25
$109.75

(9)
143
23 ct.
429
286
$32.89
12.60
$20.29

(10)
435
45 ct.
2175
1740
$195.75

$400.00
195.75
Ans. $204.25

(11)
365
65 ct.
1825
2190
$237.25

$400.00
237.00
$162.75, *Ans.*

(12)
21
3 bu.
63 bu.

63
35 ct.
315
189
Ans. $22.05

(13)

```
19 yd.        76
 4            23 ct.
76 yd.       228
            152
          $17.48, Ans.
```

(14)

```
$2000.00      5)$1836.25
  163.75       5)$367.25
$1836.25         $73.45, Ans.
```

(15)

```
4)$516.00
4)$129.00
43)$32.25($0.75, Ans.
    301
     215
     215
```

(16)

```
4|0)$9|0.00
  10)$2.25
    $0.225, Ans.
```

(17)

```
22 = 2 × 11

  $1000.00
    500.00
2)$1500.00
11)$750.00
Ans. $68.18+
```

MERCHANTS' BILLS.

(18.)

9	lb.	@	$0.32 =	$2.88	
4	"	"	1.25 =	5.00	
45	"	"	.09 =	4.05	
17	"	"	.20 =	3.40	$15.33

(19.)

22	yd.	@	$1.75 =	$38.50	
18	"	"	.15 =	2.70	
25	"	"	.65 =	16.25	
6	"	"	.18 =	1.08	$58.53

(20.)

4	lb.	@	$0.18 =	$0.72	
8	"	"	.23 =	1.84	
7	"	"	.11 =	.77	
6	"	"	.09 =	.54	
13	"	"	.35 =	4.55	
26	"	"	.12 =	3.12	$11.54

(21.)	43 yd. @ \$0.13 = \$5.59		
	28 " " .09 = 2.52		
	23 " " .23 = 5.29	\$13.40	

DRY MEASURE.

Art. 63.

(5.) 4 bu. × 4 + 2 pk. = 18 pk.: 18 pk. × 8 + 1 qt. = 145 qt.: 145 qt. × 2 = 290 pt., *Ans.*

(6.) 7 bu. × 4 + 3 pk. = 31 pk.: 31 pk. × 8 + 7 qt. = 255 qt.: 255 qt. × 2 + 1 pt. = 511 pt., *Ans.*

(7.) 3 bu. × 4 = 12 pk.: 12 pk. × 8 = 96 qt.: 96 qt. × 2 + 1 pt. = 193 pt., *Ans.*

(8.) 384 pt. ÷ 2 = 192 qt.: 192 qt. ÷ 8 = 24 pk.: 24 pk. ÷ 4 = 6 bu., *Ans.*

(9.) 47 pt. ÷ 2 = 23 qt. 1 pt.: 23 qt. ÷ 8 = 2 pk. 7 qt. *Ans.* 2 pk. 7 qt. 1 pt.

(10.) 95 pt. ÷ 2 = 47 qt. 1 pt.: 47 qt. ÷ 8 = 5 pk. 7 qt.: 5 pk. ÷ 4 = 1 bu. 1 pk. Collecting the different remainders, the *Ans.* is 1 bu. 1 pk. 7 qt. 1 pt.

(11.) 508 pt. ÷ 2 = 254 qt.: 254 qt. ÷ 8 = 31 pk. 6 qt.: 31 pk. ÷ 4 = 7 bu. 3 pk. *Ans.* 7 bu. 3 pk. 6 qt.

LIQUID MEASURE.

Art. 64.

(1.) 17 gal. × 4 × 2 = 136 pt., *Ans.*

(2.) 13 gal. × 4 × 2 × 4 = 416 gi., *Ans.*

(3.) 126 gal. × 4 × 2 = 1008 pt., *Ans.*

(4.) 1260 gal. × 4 × 2 × 4 = 40320 gi., *Ans.*

(5.) 1120 gi. ÷ 4 = 280, ÷ 2 = 140, ÷ 4 = 35 gal., *Ans.*

(6.) 1848 cu. in. $\div$ 231 = 8 gal., *Ans.*

(7.) 138138 cu. in. $\div$ 231 = 598 gal., *Ans.*

Avoirdupois Weight.

Art. 65.

(1.) 2 cwt. $\times$ 4 $\times$ 25 = 200 lb., *Ans.*

(2.) 3 cwt. $\times$ 100 = 300 lb. + 75 lb. = 375 lb., *Ans.*

(3.) 1 T. $\times$ 20 + 2 cwt. = 22 cwt. $\times$ 100 = 2200 lb., *Ans.*

(4.) 3 T. $\times$ 20 $\times$ 100 = 6000 lb. + 75 lb. = 6075 lb., *Ans.*

(5.) 4 cwt. $\times$ 100 + 44 lb. = 444 lb., *Ans.*

(6.) 5 T. $\times$ 20 $\times$ 100 + 90 lb. = 10090 lb., *Ans.*

(7.) 2 cwt. $\times$ 100 + 77 lb. = 277 lb.: 277 lb. $\times$ 16 + 12 oz. = 4444 oz., *Ans.*

(8.) 2 cwt. $\times$ 100 + 17 lb. = 217 lb.: 217 lb. $\times$ 16 + 3 oz. = 3475 oz., *Ans.*

(9.) 1 T. $\times$ 20 + 6 cwt. = 26 cwt., $\times$ 100 + 4 lb. = 2604 lb., $\times$ 16 + 2 oz. = 41666 oz., *Ans.*

(10.) 4803 lb. $\div$ 100 = 48 cwt. and 3 lb. over, *Ans.*

(11.) 22400 lb. $\div$ 100 $\div$ 20 = 11 T. and 4 cwt., *Ans.*

(12.) 2048000 $\div$ 16 = 128000 lb., $\div$ 100 = 1280 cwt., $\div$ 20 = 64 T., *Ans.*

(13.) 64546 oz. $\div$ 16 = 4034 lb. 2 oz.: 4034 $\div$ 100 = 40 cwt. 34 lb. *Ans.* 40 cwt. 34 lb. 2 oz.

(14.) 97203 oz. $\div$ 16 = 6075 lb. 3 oz.: 6075 $\div$ 100 = 60 cwt. 75 lb.: 60 $\div$ 20 = 3 T. *Ans.* 3 T. 75 lb. 3 oz.

(15.) 544272 oz. $\div$ 16 = 34017 lb., $\div$ 100 = 340 cwt. 17 lb.: 340 $\div$ 20 = 17 T. *Ans.* 17 T. 17 lb.

(16.) $52 \times 18 = 936$ lb.: $936 \div 100 = 9$ cwt. 36 lb., *Ans.*

(17.) $180 \times 75 = 13500$ lb.: $13500 \div 100 = 135$ cwt. $\div 20 = 6$ T. 15 cwt., *Ans.*

Long Measure.

Art. 66.

(1.) 2 yd. $\times 3 + 2$ ft. $= 8$ ft.: 8 ft. $\times 12 + 7$ in. $= 103$ in., *Ans.*

(2.) 7 yd. $\times 3 = 21$ ft., $\times 12 + 11$ in: $= 263$ in., *Ans.*

(3.) 12 mi. $\times 320 = 3840$ rd., *Ans.*

(4.) 7 mi. $\times 320 + 240$ rd. $= 2480$ rd., *Ans.*

(5.) 9 mi. $\times 320 + 31$ rd. $= 2911$ rd., *Ans*

(6.) 133 in. $\div 12 = 11$ ft. 1 in.: 11 ft. $\div 3 = 3$ yd. 2 ft. *Ans.* 3 yd. 2 ft. 1 in.

(7.) 181 in. $\div 12 = 15$ ft. 1 in.: 15 ft. $\div 3 = 5$ yd. *Ans.* 5 yd. 1 in.

(8.) 2240 rd. $\div 320 = 7$ mi., *Ans.*

(9.) 2200 rd. $\div 320 = 6$ mi. 280 rd., *Ans.*

(10.) 1 mi. $\times 320 \times 5\frac{1}{2} = 1760$ yd., *Ans.*

(11.) 1 mi. $\times 320 \times 5\frac{1}{2} \times 3 = 5280$ ft. *Ans.*

Square Measure.

Art. 67.

(1.) 8 sq. yd. $\times 9 \times 144 = 10368$ sq. in., *Ans.*

(2.) 4 A. $\times 160 = 640$ sq. rd., *Ans.*

(3.) 1 sq. mi. $\times 640 \times 160 = 102400$ sq. rd., *Ans.*

(4.) 2 sq. yd. $\times 9 + 3$ sq. ft. $= 21$ sq. ft.: 21 sq. ft. $\times 144 = 3024$ sq. in., *Ans.*

(5.) 5 A. $\times 160 + 100$ sq. rd. $= 900$ sq. rd., *Ans.*

(6.) 960 sq. rd. $\div$ 160 $=$ 6 A., *Ans.*

(7.) 3888 sq. in. $\div$ 144 $=$ 27 sq. ft.: 27 sq. ft. $\div$ 9 $=$ 3 sq. yd., *Ans.*

(8.) 20000 sq. rd. $\div$ 160 $=$ 125 A., *Ans.*

(9.) 515280 sq. rd. $\div$ 160 $\div$ 640 $=$ 5 sq. mi. 20 A. 80 sq. rd., *Ans.*

(10.) 4176 sq. in. $\div$ 144 $=$ 29 sq. ft.: 29 sq. ft. $\div$ 9 $=$ 3 sq. yd. 2 sq. ft., *Ans.*

Art. 68.

(2.) 16 ft. $\times$ 12 ft. $=$ 192 sq. ft., *Ans.*

(3.) 5 yd. $\times$ 4 yd. $=$ 20 sq. yd., *Ans.*

(4.) 18 ft. $\div$ 3 $=$ 6 yd.: 12 ft. $\div$ 3 $=$ 4 yd.: 21 ft. $\div$ 3 $=$ 7 yd.: 15 ft. $\div$ 3 $=$ 5 yd. 6 yd. $\times$ 4 yd. $=$ 24 sq. yd.: 7 yd. $\times$ 5 yd. $=$ 35 sq. yd.: 24 sq. yd. $+$ 35 sq. yd. $=$ 59 sq. yd., *Ans.*

(5.) 18 ft. $\times$ 14 ft. $=$ 252 sq. ft.: 252 sq. ft. $\div$ 9 $=$ 28 sq. yd., *Ans.*

(6.) 35 rd. $\times$ 32 rd. $=$ 1120 sq. rd.: 1120 sq. rd. $\div$ 160 $=$ 7 A., *Ans.*

(7.) 18 ft. $\div$ 3 $=$ 6 yd.: 15 ft. $\div$ 3 $=$ 5 yd. 5 yd. $\times$ 6 yd. $\times$ 2 $=$ 60 sq. yd.: 60 $\times$ \$1.25 $=$ \$75, *Ans.*

(8.) 21 ft. $=$ 7 yd.: 18 ft. $=$ 6 yd.; 7 yd. $\times$ 6 yd. $=$ 42 sq. yd.: 42 $\times$ \$0.17 $=$ \$7.14, *Ans.*

Art. 69.

(1.) 132 sq. ft. $\div$ 11 ft. $=$ 12 ft., *Ans.*

(2.) 30 sq. yd. $\times$ 9 $=$ 270 sq. ft.: 270 sq. ft. $\div$ 18 ft. $=$ 15 ft., *Ans.*

(3.) 9 A. × 160 = 1440 sq. rd.: 1440 sq. rd. ÷ 45 rd. = 32 rd., *Ans.*

(4.) 21 A. × 160 = 3360 sq. rd.: 3360 sq. rd. ÷ 35 rd. = 96 rd., *Ans.*

Cubic Measure.

Art. 70.

(1.) 2 cu. yd. × 27 × 1728 = 93312 cu. in., *Ans.*

(2.) 28 C. × 128 = 3584 cu. ft., *Ans.*

(3.) 34 C. × 128 × 1728 = 7520256 cu. in., *Ans.*

(4.) 1 C. × 128 × 1728 = 221184 cu. in., *Ans.*

(5.) 63936 cu. in. ÷ 1728 = 37 cu. ft.: 37 cu. ft. ÷ 27 = 1 cu. yd. 10 cu. ft., *Ans.*

(6.) 8 ft. × 5 ft. × 4 ft. = 160 cu. ft., *Ans.*

(7.) 8 yd. × 5 yd. × 2 yd. = 80 cu. yd., *Ans.*

(8.) 18 ft. × 15 ft. × 7 ft. = 1890 cu. ft.: 1890 cu. ft. ÷ 27 = 70 cu. yd., *Ans.*

(9.) 40 ft. × 12 ft. × 8 ft. = 3840 cu. ft.: 3840 cu. ft. ÷ 128 = 30 C., *Ans.*

(10.) 80 ft. × 8 ft. × 4 ft. = 2560 cu. ft.: 2560 cu. ft. ÷ 128 = 20 C.: 20 × $5.50 = $110, *Ans.*

(11.) 24 ft. = 8 yd., 15 ft. = 5 yd., 6 ft. = 2 yd.: 8 yd. × 5 yd. × 2 yd. = 80 cu. yd.: 80 × $1.25 = $100, *Ans.*

Time Measure.

Art. 71.

(1.) 2 hr. × 60 × 60 = 7200 sec., *Ans.*

(2.) 7 da. × 24 × 60 = 10080 min., *Ans.*

(3.) 1 da. × 24 + 3 hr. = 27 hr.: 27 hr. × 60 + 44 min. = 1664 min.: 1664 min. × 60 + 3 sec. = 99843 sec., *Ans.*

(4.) 9 wk. $\times$ 7 + 6 da. = 69 da.: 69 da. $\times$ 24 + 10 hr. = 1666 hr.: 1666 hr. $\times$ 60 + 40 min. = 100000 min., *Ans.*

(5.) 4 wk. $\times$ 7 + 3 da. = 31 da. : 31 da. $\times$ 24 = 744 hr. : 744 hr. $\times$ 60 + 4 min. = 44644 min., *Ans.*

(6.) 10800 sec. $\div$ 60 = 180 min.: 180 min. $\div$ 60 = 3 hr., *Ans.*

(7.) 432000 sec. $\div$ 60 = 7200 min.: 7200 min. $\div$ 60 = 120 hr.: 120 hr. $\div$ 24 = 5 da., *Ans.*

(8.) 7322 sec. $\div$ 60 = 122 min. 2 sec.: 122 min. $\div$ 60 = 2 hr. 2 min. *Ans.* 2 hr. 2 min. 2 sec.

(9.) 4323 min. $\div$ 60 = 72 hr. 3 min.: 72 hr. $\div$ 24 = 3 da. *Ans.* 3 da. 3 min.

(10.) 20280 min. $\div$ 60 = 338 hr.: 338 hr. $\div$ 24 = 14 da. 2 hr.: 14 da. $\div$ 7 = 2 wk. *Ans.* 2 wk. 2 hr.

(11.) 41761 min. $\div$ 60 = 696 hr. 1 min.: 696 hr. $\div$ 24 = 29 da.: 29 da. $\div$ 7 = 4 wk. 1 da.: 4 wk. $\div$ 4 = 1 mo. *Ans.* 1 mo. 1 da. 1 min.

Miscellaneous Tables.

Art. 73.

(1.) 5 lb. $\times$ 12 + 4 oz. = 64 oz., *Ans.*

(2.) 9 lb. $\times$ 12 + 3 oz. = 111 oz.: 111 oz. $\times$ 20 + 5 pwt. = 2225 pwt., *Ans.*

(3.) 8 lb. $\times$ 12 + 9 oz. = 105 oz.: 105 oz. $\times$ 20 + 13 pwt. = 2113 pwt.: 2113 pwt. $\times$ 24 + 17 gr. = 50729 gr., *Ans.*

(4.) 805 pwt. $\div$ 20 = 40 oz. 5 pwt.: 40 oz. $\div$ 12 = 3 lb. 4 oz. *Ans.* 3 lb. 4 oz. 5 pwt.

(5.) 12530 gr. $\div$ 24 = 522 pwt. 2 gr.: 522 pwt. $\div$ 20 = 26 oz. 2 pwt.: 26 oz. $\div$ 12 = 2 lb. 2 oz. *Ans.* 2 lb. 2 oz. 2 pwt. 2 gr.

(6.) 4 ℔. × 12 + 5 ℥ = 53 ℥: 53 × 8 × 3 × 20 + 2 gr. = 25442 gr., *Ans.*

(7.) 7 ℔. × 12 + 2 ℥ = 86 ℥: 86 ℥ × 8 = 688 ʒ: 688 ʒ × 3 + 1 ℈ = 2065 ℈: 2065 ℈ × 20 = 41300 gr., *Ans.*

(8.) 431 ʒ ÷ 8 = 53 ℥ 7 ʒ: 53 ℥ ÷ 12 = 4 ℔. 5 ℥. *Ans.* 4 lb. 5 ℥ 7 ʒ.

(9.) 975 ℈ ÷ 3 = 325 ʒ: 325 ʒ ÷ 8 = 40 ℥ 5 ʒ: 40 ℥ ÷ 12 = 3 ℔. 4 ℥. *Ans.* 3 ℔. 4 ℥ 5 ʒ.

(10.) 6321 gr. ÷ 20 = 316 ℈ 1 gr.: 316 ℈ ÷ 3 = 105 ʒ 1 ℈: 105 ʒ ÷ 8 = 13 ℥ 1 ʒ: 13 ℥ ÷ 12 = 1 ℔. 1 ℥. *Ans.* 1 ℔. 1 ℥ 1 ʒ 1 ℈ 1 gr.

(11.) 4 cong. × 8 × 16 + 7 f. ℥ × 8 = 4152 f. ʒ., *Ans.*

(12.) 5 O. × 16 + 6 f. ℥ = 86 f. ℥: 86 f. ℥ × 8 + 3 f. ʒ = 691 f. ʒ: 691 × 60 = 41460 minims, *Ans.*

(13.) 2469 f. ʒ ÷ 8 = 308 f. ℥ 5 f. ʒ: 308 ÷ 16 = 19 O. 4 f. ℥: 19 ÷ 8 = 2 cong. 3 O. *Ans.* 2 cong. 3 O. 4 f. ℥ 5 f. ʒ.

(14.) 3 yd. × 3 = 9 ft., × 12 = 108 in., × 3 = 324 barleycorns, *Ans.*

(15.) 1 ft. × 12 + 6 in. = 18 in.: 18 × 12 = 216 lines, *Ans.*

(16.) $16\frac{1}{2}$ hands × 4 = 66 in.: 66 ÷ 12 = 5 ft. 6 in., *Ans.*

(17.) 24 chains × 4 = 96 rd.: 15 chains × 4 = 60 rd.: 96 rd. × 60 rd. = 5760 sq. rd.: 5760 ÷ 160 = 36 A., *Ans.*

(18.) 267 cu. ft. × 1728 + 624 cu. in. = 462000 cu. in.: 462000 ÷ 231 = 2000 gal., *Ans.*

(19.) $8° \times 60 + 41' = 521'$: $521' \times 60 + 45'' = 31305''$, *Ans.*

(20.) $61° \times 60 + 59' = 3719'$: $3719' \times 60 + 28'' = 223168''$, *Ans.*

(21.) $915' \div 60 = 15° \ 15'$, *Ans.*

(22.) $3661'' \div 60 = 61' \ 1''$: $61' \div 60 = 1° \ 1'$. *Ans.* $1° \ 1' \ 1''$.

(23.) 6 gross $\times$ 12 = 72 doz., $\times$ 5 ct. = \$3.60, *Ans.*

(24.) 4 score $\times$ 20 + 10 yr. = 90 yr., *Ans.*

(25.) 3 bdl. $\times$ 2 = 6 rm., $\times$ 20 = 120 qr.: 120 @ 18 ct. = \$21.60, *Ans.*

(26.) 336 pp. $\div$ 2 = 168 leaves: 168 $\div$ 12 = 14 sheets, *Ans.*

(27.) 512 + 528 + 528 + 512 + 496 = 2576 pp., $\div$ 2 = 1288 leaves: 1288 $\div$ 8 = 161 sheets, $\div$ 24 = 6 qr. 17 sheets, *Ans.*

Art. 74.

(1.) 2 bu. $\times$ 4 $\times$ 8 $\times$ 2 = 128 pt.: 5 ct. $\times$ 128 = 640 ct. = \$6.40, *Ans.*

(2.) 3 bu. $\times$ 4 + 2 pk. = 14 pk.: 50 ct. $\times$ 14 = 700 ct. = \$7.00, *Ans.*

(3.) 3 pk. $\times$ 8 + 3 qt. = 27 qt.: 27 qt. $\times$ 2 = 54 pt.: 3 ct. $\times$ 54 = \$1.62, *Ans.*

(4.) \$3 = 300 ct.: 300 ct. $\div$ 15 ct. = 20 pk.: 20 pk. $\div$ 4 = 5 bu., *Ans.*

(5.) \$1.66 = 166 ct.: 166 ct. $\div$ 4 = 41 qt. and 2 ct. over, which will buy 1 pt. at 4 ct. a qt. 41 qt. $\div$ 8 = 5 pk. 1 qt.: 5 pk. $\div$ 4 = 1 bu. 1 pk. *Ans.* 1 bu. 1 pk. 1 qt. 1 pt.

Or thus: 4 ct. a qt. is 2 ct. a pt.; and 166 ct. $\div$ 2 ct. = 83 pt. = 1 bu. 1 pk. 1 qt. 1 pt., *Ans.*

(6.) 3 bu. 2 pk. = 14 pk.: 91 bu. = 364 pk.: 364 pk. $\div$ 14 pk. = 26 bags, *Ans.*

(7.) 15 lb. $\times$ 16 + 12 oz. = 252 oz.: 252 $\div$ 4 = 63, *Ans.*

(8.) 44 cwt. 52 lb. = 71232 oz.: 9 lb. 15 oz. = 159 oz.: 71232 $\div$ 159 = 448, *Ans.*

(9.) 14 cwt. 28 lb. = 1428 lb.: 1428 $\div$ 84 = 17, *Ans.*

(10.) 7 cwt. 56 lb. = 756 lb.: 756 $\div$ 12 = 63, *Ans.*

(11.) 6 cwt. 10 lb. = 9760 oz.: 3 lb. 13 oz. = 61 oz.: 9760 $\div$ 61 = 160, *Ans.*

(12.) 2 A. 125 sq. rd. = 445 sq. rd.: 20 ct. $\times$ 445 = 8900 ct. = $89, *Ans.*

(13.) 16 A. 53 sq. rd. = 2613 sq. rd.: 1 A. 41 sq. rd. = 201 sq. rd.: 2613 $\div$ 201 = 13, *Ans.*

(14.) 2 ft. $\times$ 2 ft. $\times$ 2 ft. = 8 cu. ft.: 8 cu. ft. $\times$ 1728 = 13824 cu. in., *Ans.*

(15.) 1000 oz. $\times$ 5 = 5000 oz. = 312 lb. 8 oz., *Ans.*

(16.) 1000 oz. $\times$ 128 = 128000 oz. = 4 T., *Ans.*

(17.) 2 C. $\times$ 128 = 256 cu. ft.: 950 oz. $\times$ 256 = 243200 oz. = 7 T. 12 cwt., *Ans.*

(18.) 63 gal. $\times$ 4 $\times$ 2 = 504 pt.: 20 ct. $\times$ 504 = 10080 ct. = $100.80, *Ans.*

(19.) 31 gal. 2 qt. = 126 qt.: 126 qt. $\times$ 5 = 630 qt.: 10 ct. $\times$ 630 = 6300 ct. = $63, *Ans.*

(20.) $2 = 200 ct.: 200 $\div$ 5 = 40 pt.: 40 pt. = 5 gal., *Ans.*

(21.) 63 gal. = 504 pt.: 3 qt. 1 pt. = 7 pt.: 7 pt. $\times$ 12 = 84 pt. in 1 doz. bottles: 504 $\div$ 84 = 6 doz., *Ans.*

(22.) 4 gal. 3 qt. 1 pt. = 39 pt.: 58 gal. 2 qt. = 468 pt.: 468 $\div$ 39 = 12, *Ans.*

(23.) 1 da. = 1440 min.: 70 beats × 1440 = 100800 beats, *Ans.*

(24.) 1876 is a leap year, because it is exactly divisible by 4; hence, February has 29 days: 29 days = 2505600 seconds, *Ans.*

(25.) 3 wk. 2 da. 3 hr. = 555 hr.: 8 mi. × 555 = 4440 mi., *Ans.*

(26.) A peck is $\frac{1}{4}$ bushel, and will, therefore, cost $\frac{1}{4}$ of 44 ct. = 11 ct. per day: 365 × 11 = $40.15, *Ans.*

(27.) 40 bbl. × 196 = 7840 lb. The gain equals 5 ct. — 3 ct., or 2 ct., a pound. 7840 × 2 ct. = $156.80, *Ans.*

Art. 75.

(4.) 17 bu. 3 pk. 7 qt., *Ans.*

(5.) 26 bu. 1 qt. 1 pt., *Ans.*

(6.) 24 qt., *Ans.*

(7.) 128 gal. 3 qt. 1 pt. 3 gi., *Ans.*

(8.) 79 T. 15 cwt. 48 lb. 6 oz., *Ans.*

(9.) 57 cwt. 51 lb. 7 oz., *Ans.*

(10.) 111 mi. 44 rd., *Ans.*

(11.) 14 yd. 4 in., *Ans.*

(12.) 299 A. 150 sq. rd., *Ans.*

(13.) 51 sq. yd. 4 sq. ft. 73 sq. in., *Ans.*

(14.) 49 C. 58 cu. ft. 519 cu. in., *Ans.*

(15.) 143 cu. yd. 2 cu. ft. 990 cu. in., *Ans.*

(16.) 50 da. 3 hr. 12 min. 28 sec., *Ans.*

(17.) 8 mo. 4 da. 8 hr. 49 min. 35 sec., *Ans.*

(18)

bu.	pk.
21	3
14	1
23	2
18	1
22	1
100	0

(19)

bu.	pk.
200	3
143	1
400	3
255	1
1000	0

(20)

cwt.	lb.
8	36
4	64
5	19
7	75
7	84
33	78

(21)

lb.	oz.
13	11
17	13
14	14
16	0
19	7
17	9
99	6

(22)

mi.	rd.
104	50
95	270
200	0

(23)

A.	sq. rd.
186	134
286	17
113	89
586	80

(24)

sq. yd.	sq. ft.	sq. in.
17	3	119
18	0	141
23	7	0
29	5	116
88	8	88

(25)

C.	cu. ft.
7	78
16	24
35	127
29	10
88	111

(26)

hhd.	gal.	qt.	pt.
	4642	3	1
	945	0	0
	1707	0	1
	10206	1	0
277	50	1	0

Art. 76.

(4.) 3 gal. 3 qt. 1 pt., *Ans.*

(5.) 19 gal. 1 qt. 1 pt. 3 gi., *Ans.*

(6.) 3 T. 18 cwt. 75 lb., *Ans.*

(7.) 11 T. 42 lb. 15 oz., *Ans.*

(8.) 6 mi. 282 rd., *Ans.*

(9.) 1 yd. 2 ft. 11 in., *Ans.*

(10.) 249 A. 153 sq. rd., *Ans.*

(11.) 2 sq. yd. 8 sq. ft. 104 sq. in., *Ans.*

(12.) 8 C. 125 cu. ft., *Ans.*

(13.) 8 cu. yd. 18 cu. ft. 1727 cu. in., *Ans.*

(14.) 51 min. 42 sec., *Ans.*

(15.) 55 da. 5 hr. 55 min. 55 sec., *Ans.*

(16)

bu.	pk.	qt.
4	0	0
2	1	1
1	2	7

(17)

bu.	pk.	qt.	pt.
100	0	0	0
24	0	0	1
75	3	7	1

(18)

lb.	oz.
46	4
19	8
26	12

(19)

cwt.	lb.
32	66
8	67
23	99

(20)

mi.	rd.
24899	0
100	41
24798	279

(21)

A.	sq. rd.
146	80
86	94
59	146

(22)

C.	cu. ft.
8	50
3	75
4	103

(23)

gal.	qt.	pt.	gi.
63	0	0	0
51	1	0	2
11	2	1	2

(24)

da.	hr.	min.	sec.
5	10	27	15
2	4	13	29
3	6	13	46

Art. 77.

(2)

yr.	mo.	da.
1876	9	1
1776	7	4
100	1	27

(3)

yr.	mo.	da.
1191	7	12
1099	7	15
91	11	27

(4)

yr.	mo.	da.
1587	2	8
1215	6	15
371	7	23

(5)

yr.	mo.	da.
1688	11	5
1066	10	14
622	0	21

(6)

yr.	mo.	da.
1815	6	18
1805	12	2
9	6	16

Art. 78.

(2)

	da.
Mar.	14
Apr.	30
May	31
June	30
July	31
Aug.	31
Sept.	12
	179 da.

(3)

da.
12
31
20
63 da.

(4)

da.
25
31
30
7
93 da.

(5)

da.
19
30
31
31
28
31
30
25
225 da.

(6)

da.
11
30
31
30
31
31
29
8
201 da.

Art. 79.

(2)

bu.	pk.	qt.	pt.
2	1	1	1
			6
13	3	1	0

(3)

bu.	pk.	qt.
2	2	2
		9
23	0	2

(4)

bu.	pk.	qt.	pt.
4	3	3	1
			12
58	1	2	0

(5)

T.	cwt.	lb.
	8	62
		9
3	17	58

(6)

T.	cwt.	lb.
	10	89
		7
3	16	23

(7)

mi.	rd.
208	176
	15
3128	80

(8)

cu. yd.	cu. ft.	cu. in.
23	9	228
		12
280	1	1008

(9)

T.	cwt.	lb.
	16	74
		119
99	12	6

(10)

gal.	qt.	pt.
47	3	1
		59
2824	2	1

(11)

mi.	rd.
27	155
	31
852	5

(12)

C.	cu. ft.
7	98
	17
132	2

(13)

mo.	wk.	da.	hr.	min.	sec.
	2	4	13	48	39
					75
49	3	0	3	48	45

(14)

cwt.	lb.
10	84
	75
813	0

813 cwt. = 81300 lb.
8 ct. − 6 ct. = 2 ct., gain on 1 lb.
2 ct. × 81300 = 162600 ct. = $1626, *Ans.*

(15)

4 cwt. 85 lb. = 485 lb.
485 lb. × 425 = 206125 lb.
206125 × 13 ct. = $26796.25
$26796.25 − $24735 = $2061.25, *Ans.*

DIVISION.

Art. 80.

(4)

	bu.	pk.	qt.	pt.
5)	67	3	4	1
	13	2	2	1

(5)

	cwt.	lb.	oz.
11)	35	44	12
	3	22	4

(6)

	mi.	rd.
7)	39	288
	5	224

(7)

```
    A. sq. rd.
16)69   64
   ---------
     4  54
```

(8)

```
      bu.   pk.  qt.
10)490    2    4
   ---------------
10)49     0    2
   ---------------
Ans. 4    3    5
```

(9)

```
     lb.   oz.
 5)265   10
  -----------
10)53     2
  -----------
Ans. 5    5
```

(10)

```
    T.  cwt.
17)45   18
   --------
    2   14
```

(11)

```
 dr.   hr. min. sec.
6)114   22   45   18
 --------------------
9)19     3   47   33
 --------------------
Ans. 2   3    5   17
```

(12)

```
                   lb.      oz.
                   27       13
  10 cwt. =      1000
                 ----
              23)1027(44 lb.
                  92
                 ----
                 107
                  92
                 ----
                  15
15 × 16 + 13 oz. ÷ 23 = 11 oz.
       Ans. 44 lb. 11 oz.
```

(13)

```
     bu.  pk. qt. bu. pk. qt.
78)309    2   2 ( 3   3   7, Ans.
   234
   ---
    75
     4
   ---
78)302(3 pk.
   234
   ---
    68
     8
   ---
78)546(7 qt.
   546
```

(14)

```
     gal.   qt.   pt.   gi.
63)127      3     1     3
   ---
gal. 2 . . 1
           4  qt.  qt.
          ---
          4+3=7
 pt.                  2  pt.  pt.
                   ------
 15                 14+1=15
  4
 ---
 60+3gi.=63 gi.
           63÷63=1 gi
     Ans. 2 gal. 1 gi.
```

(15)

```
   mi.   rd.  mi.  rd.
319)788  169 ( 2  151, Ans.
    638
    ---
    150×320=48000 rd.
               169
             -----
       319)48169(151 rd.
           319
           ----
           1626
           1595
           ----
             319
             319
             ---
```

(16)

```
   A.  sq. rd.
  104  117
   87   78
  --------
  191  195
   40   40
  --------
3)151  155
  --------
   50 . . 1=160 sq. rd.
```

160 sq. rd.+155 sq. rd.=315 sq. rd.
315 sq. rd.÷3=105 sq. rd.
Ans. 50 A. 105 sq. rd.

(17)

```
  bu.   pk.
 5000   3
 7245   2
 ---------
12245   5
 8022   1
 ---------
 4223   4=4224 bu.
 4224 ÷ 8=528 bu., Ans.
```

(18)

```
  A.  sq. rd.
   4  80
 160
 -------
 640+80=720 sq. rd.
720 sq. rd.×6=4320 sq. rd.
54)4320(80 sq. rd. in each lot.
   432
   ---
     0
```

80×$5=$400, *Ans.*

(19)

```
 lb.   oz.
  35    9
  75   14
  85   15
 ---------
 195   38
 186   14
 ---------
   9   24
        8
 ---------
  72  192=84 lb.
```

84 lb.÷64=1 lb. 5 oz., *Ans.*

Art. 81.

(1)

$$15\overline{)18^\circ \quad 25' \quad 30''}$$
1 hr. 13 min. 42 sec.

(2)

30° ÷ 15 = 2. *Ans.* 2 hr.

(3)

$$15\overline{)71^\circ \quad 4' \quad 0''}$$
4 hr. 44 min. 16 sec.

(4)

$$15\overline{)10^\circ \quad 35' \quad 0''}$$
0 hr. 42 min. 20 sec.

(5)

	min.	sec.
	37	20
		15
9°	20′	0″

(6)

hr.	min.	sec.
1	4	56
		15
16°	14′	0″

(7)

hr.	min.	sec.
5	8	4
		15
77°	1′	0″

Art. 82.

(8)

	hr.	min.	sec.
Time at C.	12	0	0
Add diff.		37	20
	12	37	20

(See Ex. 5, Art. 81.)

(9)

	hr.	min.	sec.	
Time at N. Y.	11	0	0	A. M.
30°=	2	0	0	to be added.
	1	0	0	P. M.

(10)

	hr.	min.	sec.
Time at Ph.	12	0	0
Subtr. diff.		37	20
	11	22	40 A. M.

(11)

	hr.	min.	sec.
Time at N. Y.	11	0	0
Subtr. diff.	1	4	56
	9	55	4 A. M.

(See Ex. 6, Art. 81.)

(12)

124°−80° 42′=43° 18′: 43° 18′÷15=2 hr. 53 min. 12 sec.

	hr.	min.	sec.
Time at W.	1	0	0
Subtr. diff.	2	53	12
Ans.	10	6	48 A. M.

NOTE.—In performing the subtraction, we can not take 3 hr. from 1 hr., but 1 P. M. is the 13th hour from midnight, from which, after taking 3 hr., the remainder is the 10th hr. from midnight, or 10 A. M.

FACTORING.

Arts. 87 and 88.

NOTE.—The principles and processes of factoring are so simple, and are so fully explained in the Arithmetic, that it seems unnecessary to give any solutions here.

Art. 89.

(2)

2)16	2)24	2)40
2)8	2)12	2)20
2)4	2)6	2)10
2)2	3	5
1		

2×2×2=8, G. C. D.

(3)

2)24	2)36	2)60
2)12	2)18	2)30
3)6	3)9	3)15
2)2	3)3	5
1	1	

2×2×3=12, G. C. D.

(4)

2)36	54	90
3)18	27	45
3)6	9	15
2	3	5

2×3×3=18, G. C. D.

(5)

2)40	60	100
2)20	30	50
5)10	15	25
2	3	5

2×2×5=20, G. C. D.

(6)

3)54	81	108
3)18	27	36
3)6	9	12
2	3	4

3×3×3=27, G. C. D.

(7)

2)60	90	120
3)30	45	60
5)10	15	20
2	3	4

2×3×5=30, G. C. D.

(8)

2)32	48	80	112
2)16	24	40	56
2)8	12	20	28
2)4	6	10	14
2	3	5	7

2×2×2×2=16, G. C. D.

(9)

2)48	72	96	120
2)24	36	48	60
2)12	18	24	30
3)6	9	12	15
2	3	4	5

2×2×2×3=24, G. C. D.

(10)

2)72	108	144	180
2)36	54	72	90
3)18	27	36	45
3)6	9	12	15
2	3	4	5

2×2×3×3=36, G. C. D.

(11)

(By 2d method.)

62)93(1
62
31)62(2
62
31 = G. C. D.

(12)

78)130(1
78
52)78(1
52
26)52(2
52
26 = G. C. D.

(13)

161)253(1
161
92)161(1
92
69)92(1
69
23)69(3
69
23 = G. C. D.

(14)

247)323(1
247
76)247(3
228
19)76(4
76
19 = G. C. D.

(16)

2145)3471(1
2145
1326)2145(1
1326
819)1326(1
819
507)819(1
507
312)507(1
312
195)312(1
195
117)195(1
117
78)117(1
78
39)78(2
78

39 = G. C. D.

(15)

391)697(1
391
306)391(1
306
85)306(3
255
51)85(1
51
34)51(1
34
17)34(2
34

17 = G. C. D.

(17)

16571)38363(2
33142
5221)16571(3
15663
908)5221(5
4540
681)908(1
681
227)681(3
681

227 = G. C. D.

(18)

72)120(1
72
48)72(1
48
24)48(2
48

24)132(5
120
12)24(2
24

12 = G. C. D.

(19)

75)125(1
75
50)75(1
50
25)50(2
50

25)165(6
150
15)25(1
15
10)15(1
10
5)10(2
10

5 = G. C. D.

(20)

2)64	96	112	136
2)32	48	56	68
2)16	24	28	34
8	12	14	17

$2\times2\times2=8$, G. C. D.

Art. 90.

(2)

2)4	6	8
2)2	3	4
	2)3	2
	3	

$2\times2\times2\times3=24$, L. C. M.

(3)

3)6	9	12
2)2	3	4
	3	2

$3\times2\times3\times2=36$, L. C. M.

(4)

2)4	8	10
2)2	4	5
	2	5

$2\times2\times2\times5=40$, L. C. M.

(5)

5)6	10	15
3)6	2	3
2)2	2	

$5\times3\times2=30$, L. C. M.

(6)

3)6	8	9	12
2)2	8	3	4
	2)4	3	2
	2	3	

$3\times2\times2\times2\times3=72$, L. C. M.

(7)

5)10	12	15	20
2)2	12	3	4
	3)6	3	2
	2)2		2

$5\times2\times3\times2=60$, L. C. M.

(8)

3)9	15	18	30
5)3	5	6	10
3)3		6	2
		2)2	2

$3\times5\times3\times2=90$, L. C. M.

(9)

3)12	18	27	36
3)4	6	9	12
2)4	2	3	4
2)2		3	2
		3	

$3\times3\times2\times2\times3=108$, L. C. M.

(10)

5)15 25 30 50
5)3 5 6 10
2)3 2 2
3

$5\times5\times2\times3=150$, L. C. M.

(11)

7)14 21 30 35
5)2 3 30 5
3)2 3 6
2)2 2

$7\times5\times3\times2=210$, L. C. M.

(12)

7)15 20 21 28
5)15 20 3 4
3)3 4 3 4
2)4 4
2)2 2

$7\times5\times3\times2\times2=420$, L. C. M.

(13)

5)20 24 28 30
3)4 24 28 6
2)4 8 28 2
2)2 4 14
2 7

$5\times3\times2\times2\times2\times7=840$, L. C. M.

(14)

7)45 30 35 42
5)45 30 5 6
3)9 6 6
2)3 2 2
3

$7\times5\times3\times2\times3=630$, L. C. M.

(15)

5)36 40 45 50
3)36 8 9 10
2)12 8 3 10
2)6 4 3 5
3)3 2 3 5
2 5

$5\times3\times2\times2\times3\times2\times5=1800$, L. C. M.

(16)

7)42 56 63
3)6 8 9
2)2 8 3
2)4 3
2 3

$7\times3\times2\times2\times2\times3=504$, L. C. M.

(17)

13)78 104 117
3)6 8 9
2)2 8 3
2)4 3
2 3

$13\times3\times2\times2\times2\times3=936$, L. C. M.

(18)

5)	125	150	200
5)	25	30	40
2)	5	6	8
2)	5	3	4
	5	3	2

5×5×2×2×5×3×2=3000, L. C. M.

(19)

5)	10	24	25	32	45
3)	2	24	5	32	9
2)	2	8	5	32	3
2)		4	5	16	3
2)		2	5	8	3
2)			5	4	3
			5	2	3

5×3×2×2×2×2×5×2×3=7200, L. C. M.

(20)

3)	2	3	4	5	6	7	8	9
2)	2		4	5	2	7	8	3
2)			2	5		7	4	3
				5		7	2	3

3×2×2×5×7×2×3=2520, L. C. M.

(21)

3)	16	27	42	108
2)	16	9	14	36
3)	8	9	7	18
3)	8	3	7	6
2)	8		7	2
2)	4		7	
	2		7	

3×2×3×3×2×2×2×7=3024, L. C. M.

(22)

13)	13	29	52	87
29)		29	4	87
2)			4	3
			2	3

13×29×2×2×3=4524, L. C. M.

(23)

5)	120	360	144	720	72
3)	24	72	144	144	72
3)	8	24	48	48	24
2)	8	8	16	16	8
2)	4	4	8	8	4
2)	2	2	4	4	2
		2)	2	2	

$5\times3\times3\times2\times2\times2\times2=720$, L. C. M.

CANCELLATION.

Art. 91.

(4.) $\dfrac{\cancel{13}\times4}{\cancel{13}}=4$, *Ans.*

(5.) $\dfrac{17\times\overset{3}{\cancel{18}}}{\cancel{6}}=51$, *Ans.*

(6.) $\dfrac{15\times\overset{2}{\cancel{8}}}{\cancel{4}}=30$, *Ans.*

(7.) $\dfrac{\overset{3}{\cancel{24}}\times4}{\cancel{8}}=12$, *Ans.*

(8.) $\dfrac{37\times\overset{3}{\cancel{15}}}{\cancel{5}}=111$, *Ans.*

(9.) $\dfrac{\overset{6}{\cancel{36}}\times\overset{\cancel{5}}{\cancel{40}}}{\underset{\cancel{5}}{\cancel{30}}\times\cancel{8}}=6$, *Ans.*

(10.) $\dfrac{\overset{12}{\cancel{36}}\times\cancel{5}}{\underset{\cancel{3}}{\cancel{15}}}=12$, *Ans.*

(11.) $\dfrac{\overset{2}{\cancel{42}}\times\overset{5}{\cancel{25}}\times\overset{6}{\cancel{18}}}{\cancel{21}\times\underset{\cancel{3}}{\cancel{15}}}=60$, *Ans.*

(12.) $\dfrac{23\times\overset{2}{\cancel{10}}}{\cancel{5}}=46$, *Ans.*

(13.) $\dfrac{\overset{3}{\cancel{15}}\times\overset{2}{\cancel{14}}}{\underset{\cancel{7}}{\cancel{35}}}=6$, *Ans.*

(14.)
$$\frac{\overset{\cancel{3}}{\cancel{21}}\times 11\times \overset{3}{\cancel{6}}\times \overset{\cancel{2}}{\cancel{26}}}{\cancel{13}\times \cancel{3}\times \underset{\cancel{2}}{\cancel{14}}\times \cancel{2}}=33,\ \textit{Ans.}$$

(15.)
$$\frac{\overset{7}{\cancel{21}}\times \overset{\cancel{3}}{\cancel{15}}\times \overset{\cancel{3}}{\cancel{33}}\times \overset{\cancel{2}}{\cancel{8}}\times \overset{7}{\cancel{14}}\times \cancel{17}}{\underset{\cancel{4}}{\cancel{20}}\times \underset{\cancel{17}}{\cancel{34}}\times \underset{\cancel{2}}{\cancel{22}}\times \underset{\underset{\cancel{3}}{\cancel{9}}}{\cancel{27}}}=49,\ \textit{Ans.}$$

(16.)
$$\frac{\overset{\cancel{3}}{\cancel{21}}\times \overset{19}{\cancel{95}}\times \overset{2}{\cancel{6}}}{\underset{\cancel{5}}{\cancel{35}}\times \underset{\cancel{3}}{\cancel{9}}}=38,\ \textit{Ans.}$$

(17.)
$$\frac{\overset{5}{\cancel{35}}\times \overset{\cancel{3}}{\cancel{39}}\times \overset{\overset{\cancel{2}}{\cancel{4}}}{\cancel{40}}}{\underset{\cancel{2}}{\cancel{26}}\times \underset{\cancel{3}}{\cancel{30}}\times \underset{\underset{3}{\cancel{6}}}{\cancel{42}}}=\frac{5}{3}=1\tfrac{2}{3},\ \textit{Ans.}$$

(18.)
$$\frac{\overset{13}{\cancel{26}}\times \overset{11}{\cancel{33}}\times \overset{7}{\cancel{35}}}{\underset{2}{\cancel{4}}\times \underset{3}{\cancel{9}}\times \underset{5}{\cancel{25}}}=\frac{13\times 11\times 7}{2\times 3\times 5}=\frac{1001}{30}=33\tfrac{11}{30},\ \textit{Ans.}$$

(19.)
$$\frac{\cancel{6}\times 9\times \overset{3}{\cancel{15}}\times \overset{3}{\cancel{21}}}{4\times \cancel{6}\times \underset{2}{\cancel{10}}\times \underset{2}{\cancel{14}}}=5\tfrac{1}{16},\ \textit{Ans.}$$

(20.)
$$\frac{\overset{\cancel{3}}{\cancel{21}}\times \overset{\overset{2}{\cancel{4}}}{\cancel{24}}\times \overset{7}{\cancel{28}}\times \overset{7}{\cancel{35}}}{\underset{\cancel{2}}{\cancel{14}}\times \underset{\cancel{3}}{\cancel{18}}\times \underset{5}{\cancel{20}}\times \underset{5}{\cancel{25}}}=\frac{98}{25}=3\tfrac{23}{25},\ \textit{Ans.}$$

FRACTIONS.

Art. 103.

(8.) 4 times $\frac{7}{7} = \frac{28}{7}$, or $\frac{7}{7} \times 4 = \frac{28}{7}$.

(9.) 8 times $\frac{9}{9} = \frac{72}{9}$, or $\frac{9}{9} \times 8 = \frac{72}{9}$.

(10.) 19 times $\frac{13}{13} = \frac{247}{13}$, or $\frac{13}{13} \times 19 = \frac{247}{13}$.

(11.) $\frac{20}{20} \times 25 = \frac{500}{20}$.

(12.) $\frac{23}{23} \times 37 = \frac{851}{23}$.

Art. 104.

(2.) $\frac{2}{2} \times 4 + \frac{1}{2} = \frac{9}{2}$.

(3.) $\frac{3}{3} \times 2 + \frac{1}{3} = \frac{7}{3}$.

(8.) $\frac{6}{6} \times 15 + \frac{5}{6} = \frac{95}{6}$.

(9.) $\frac{24}{24} \times 26 + \frac{13}{24} = \frac{637}{24}$.

(12.) $\frac{583}{583} \times 21 + \frac{117}{583} = \frac{12360}{583}$.

(14.) $\frac{71}{71} \times 14 + \frac{6}{71} = \frac{1000}{71}$.

Art. 105.

(3)
3)6
2

(4)
4)12
3

(5)
4)15
$3\frac{3}{4}$

(6)
5)17
$3\frac{2}{5}$

(7)
7)19
$2\frac{5}{7}$

(8)
10)23
$2\frac{3}{10}$

(13)
24)611($25\frac{11}{24}$
48
131
120
11

(14)
75)3000(40
300
0

(15)
25)775(31
75
25
25

(16)
12)171
$14\frac{3}{12}$

(17)
11)509($46\frac{3}{11}$
44
69
66
3

(18)
298)6437($21\frac{178}{298}$
596
477
298
179

(19)
125)7536($60\frac{36}{125}$
750
36

(20)
19)3781(199
19
188
171
171
171

(21)
101)1325($13\frac{12}{101}$
101
315
303
12

Art. 106.

(2.) $\frac{1}{2} \times \frac{2}{2} = \frac{2}{4}$, *Ans.*

(3.) $\frac{2}{3} \times \frac{2}{2} = \frac{4}{6}$, *Ans.*

(4.) $\frac{3}{4} \times \frac{3}{3} = \frac{9}{12}$, *Ans.*

(5.) $\frac{5}{6} \times \frac{4}{4} = \frac{20}{24}$, *Ans.*

(6.) $\frac{5}{7} \times \frac{4}{4} = \frac{20}{28}$, *Ans.*

(7.) $\frac{4}{21} \times \frac{4}{4} = \frac{16}{84}$, *Ans.*

(8.) $\frac{7}{8} \times \frac{9}{9} = \frac{63}{72}$, *Ans.*

(9.) $\frac{3}{5} \times \frac{12}{12} = \frac{36}{60}$, *Ans.*

(10.) $\frac{9}{10} \times \frac{10}{10} = \frac{90}{100}$, *Ans.*

(11.) 2∅)72∅; $\frac{9}{20} \times \frac{36}{36} = \frac{324}{720}$, *Ans.*
36

(12.) 14)2016(144; $\frac{13}{24} \times \frac{144}{144} = \frac{1872}{2016}$, *Ans.*
14
61
56
56
56

(13.) 43)1935(45; $\frac{22}{43} \times \frac{45}{45} = \frac{990}{1935}$, *Ans.*

172
215
215

(14.) 41)8118(198; $\frac{35}{41} \times \frac{198}{198} = \frac{6930}{8118}$, *Ans.*

41
401
369
328
328

(15.) 17)5134(302; $\frac{16}{17} \times \frac{302}{302} = \frac{4832}{5134}$, *Ans.*

51
34
34

(16.) 81)23328(288; $\frac{77}{81} \times \frac{288}{288} = \frac{22176}{23328}$, *Ans.*

162
712
648
648
648

(17.) 21)2541(121; $\frac{13}{21} \times \frac{121}{121} = \frac{1573}{2541}$, *Ans.*

21
44
42
21
21

Art. 107.

(2.) The G. C. D. of 18 and 30 is 6: $6)\frac{18}{30}=\frac{3}{5}$, *Ans.*

(3.) $10)\frac{60}{90}=\frac{6}{9}$: $3)\frac{6}{9}=\frac{2}{3}$, *Ans.*

(4.) G. C. D. of 12 and 18 $=6$: $6)\frac{12}{18}=\frac{2}{3}$, *Ans.*

(5.) $5)\frac{30}{45}=\frac{6}{9}$: $3)\frac{6}{9}=\frac{2}{3}$, *Ans.*

(6.) G. C. D. of 60 and 150 $=30$: $30)\frac{60}{150}=\frac{2}{5}$, *Ans.*

(7.) G. C. D. of 42 and 70 $=14$: $14)\frac{42}{70}=\frac{3}{5}$, *Ans.*

(8.) G. C. D. of 96 and 112 $=16$: $16)\frac{96}{112}=\frac{6}{7}$, *Ans.*

(9.) $5)\frac{60}{125}=\frac{12}{25}$, *Ans.*

(10.) $2)\frac{126}{198}=\frac{63}{99}$: $9)\frac{63}{99}=\frac{7}{11}$, *Ans.*

(11.) $2)\frac{182}{196}=\frac{91}{98}$: $7)\frac{91}{98}=\frac{13}{14}$, *Ans.*

(12.) $5)\frac{615}{915}=\frac{123}{183}$: $3)\frac{123}{183}=\frac{41}{61}$, *Ans.*

(13.) G. C. D. of 873 and 1067 $=97$: $97)\frac{873}{1067}=\frac{9}{11}$, *Ans.*

(14.) G. C. D. of 777 and 1998 $=111$: $111)\frac{777}{1998}=\frac{7}{18}$, *Ans.*

(15.) G. C. D. of 909 and 2323 $=101$: $101)\frac{909}{2323}=\frac{9}{23}$, *Ans.*

(16.) $\frac{391}{667}$: G. C. D. $=23$: $23)\frac{391}{667}=\frac{17}{29}$, *Ans.*

(17.) $\frac{585}{1287}$: G. C. D. $=117$: $117)\frac{585}{1287}=\frac{5}{11}$, *Ans.*

(18.) $\frac{796}{14129}$: G. C. D. $=199$: $199)\frac{796}{14129}=\frac{4}{71}$, *Ans.*

(19.) $\frac{1457}{5921}$: G. C. D. $=31$: $31)\frac{1457}{5921}=\frac{47}{191}$, *Ans.*

(20.) $5)\frac{6465}{7335}=\frac{1293}{1467}$, $\div 3=\frac{431}{489}$, *Ans.*

Art. 108.

(2.) 2)2 3 4

1 3 2; $2\times 3\times 2=12$, L. C. Denominator. Each must be changed to twelfths. If there are $\frac{12}{12}$ in 1, in $\frac{1}{2}$ there are $\frac{1}{2}$ of $\frac{12}{12}=\frac{6}{12}$: $\frac{1}{3}$ of $\frac{12}{12}=\frac{4}{12}$, and $\frac{2}{3}=\frac{8}{12}$: $\frac{1}{4}$ of $\frac{12}{12}=\frac{3}{12}$, and $\frac{3}{4}=\frac{9}{12}$.

(3.) L. C. M. of 3, 6, and 9 is 18; $\frac{1}{3}=\frac{6}{18}$, and $\frac{2}{3}=\frac{12}{18}$: $\frac{1}{6}=\frac{3}{18}$, and $\frac{5}{6}=\frac{15}{18}$: $\frac{1}{9}=\frac{2}{18}$, and $\frac{7}{9}=\frac{14}{18}$.

(4.) The L. C. M. of 2, 4, and 5 = 20; $\frac{1}{2}=\frac{10}{20}$: $\frac{1}{4}=\frac{5}{20}$, and $\frac{3}{4}=\frac{15}{20}$: $\frac{1}{5}=\frac{4}{20}$, and $\frac{4}{5}=\frac{16}{20}$.

(5.) L. C. M. of 8, 5, and 10 = 40: $\frac{1}{8}=\frac{1}{8}$ of $\frac{40}{40}=\frac{5}{40}$, and $\frac{3}{8}=\frac{15}{40}$: $\frac{1}{5}=\frac{1}{5}$ of $\frac{40}{40}=\frac{8}{40}$, and $\frac{4}{5}=\frac{32}{40}$: $\frac{1}{10}=\frac{1}{10}$ of $\frac{40}{40}=\frac{4}{40}$, and $\frac{9}{10}=\frac{36}{40}$.

(6.) The L. C. M. of 3, 4, and 8 is 24; $\frac{1}{3}=\frac{8}{24}$, and $\frac{2}{3}=\frac{16}{24}$: $\frac{1}{4}=\frac{6}{24}$, and $\frac{3}{4}=\frac{18}{24}$: $\frac{1}{8}=\frac{3}{24}$, and $\frac{7}{8}=\frac{21}{24}$.

(7.) L. C. M. of 4, 8, and 9 = 72; $\frac{1}{4}=\frac{18}{72}$, and $\frac{3}{4}=\frac{54}{72}$: $\frac{1}{8}=\frac{9}{72}$, and $\frac{5}{8}=\frac{45}{72}$: $\frac{1}{9}=\frac{8}{72}$, and $\frac{5}{9}=\frac{40}{72}$.

(12.) L. C. M. of 3, 5, 7, and 8 = 840; $\frac{1}{3}=\frac{280}{840}$, and $\frac{2}{3}=\frac{560}{840}$: $\frac{1}{5}=\frac{168}{840}$, and $\frac{2}{5}=\frac{336}{840}$: $\frac{1}{7}=\frac{120}{840}$, and $\frac{3}{7}=\frac{360}{840}$: $\frac{1}{8}=\frac{105}{840}$, and $\frac{5}{8}=\frac{525}{840}$.

(13.) First reduce $\frac{9}{21}$ to lowest terms $=\frac{3}{7}$. L. C. M. of 7, 14, 7, and 28 is 28; $\frac{1}{7}=\frac{4}{28}$, and $\frac{2}{7}=\frac{8}{28}$: $\frac{1}{14}=\frac{2}{28}$, and $\frac{5}{14}=\frac{10}{28}$: $\frac{1}{7}=\frac{4}{28}$, and $\frac{3}{7}=\frac{12}{28}$: $\frac{11}{28}$ is already reduced.

(14.) $\frac{6}{9}=\frac{2}{3}$: $\frac{15}{18}=\frac{5}{6}$; the L. C. M. of 5, 4, 3, and 6 is 60; $\frac{1}{5}=\frac{12}{60}$, and $\frac{2}{5}=\frac{24}{60}$: $\frac{1}{4}=\frac{15}{60}$, and $\frac{3}{4}=\frac{45}{60}$: $\frac{1}{3}=\frac{20}{60}$, and $\frac{2}{3}=\frac{40}{60}$: $\frac{1}{6}=\frac{10}{60}$, and $\frac{5}{6}=\frac{50}{60}$.

(15.) The L. C. M. of 4, 9, and 12 = 36; $1=\frac{36}{36}$, and $2=\frac{72}{36}$: $\frac{1}{4}=\frac{9}{36}$, and $\frac{3}{4}=\frac{27}{36}$: $\frac{1}{9}=\frac{4}{36}$, and $\frac{5}{9}=\frac{20}{36}$: $\frac{1}{12}=\frac{3}{36}$, and $\frac{7}{12}=\frac{21}{36}$.

(16.) $2\frac{2}{3}=\frac{8}{3}$: $5\frac{5}{6}=\frac{35}{6}$; L. C. M. of 3, 5, and 6 is 30; $\frac{1}{3}=\frac{10}{30}$, and $\frac{8}{3}=\frac{80}{30}$: $\frac{1}{5}=\frac{6}{30}$, and $\frac{3}{5}=\frac{18}{30}$: $1=\frac{30}{30}$, and $4=\frac{120}{30}$: $\frac{1}{6}=\frac{5}{30}$, and $\frac{35}{6}=\frac{175}{30}$.

(17.) $2\frac{1}{2}=\frac{5}{2}$: $3\frac{1}{3}=\frac{10}{3}$: $4\frac{1}{4}=\frac{17}{4}$; L. C. M. of 2, 3, and 4 is 12; $\frac{1}{2}=\frac{6}{12}$, and $\frac{5}{2}=\frac{30}{12}$: $\frac{1}{3}=\frac{4}{12}$, and $\frac{10}{3}=\frac{40}{12}$: $\frac{1}{4}=\frac{3}{12}$, and $\frac{17}{4}=\frac{51}{12}$: $1=\frac{12}{12}$, and $5=\frac{60}{12}$.

(18.) L. C. M. of 16, 18, 24, 36, and 48 is 144; $\frac{1}{16}=\frac{9}{144}$, and $\frac{7}{16}=\frac{63}{144}$: $\frac{1}{18}=\frac{8}{144}$, and $\frac{11}{18}=\frac{88}{144}$: $\frac{1}{24}=\frac{6}{144}$, and $\frac{17}{24}=\frac{102}{144}$: $\frac{1}{36}=\frac{4}{144}$, and $\frac{19}{36}=\frac{76}{144}$: $\frac{1}{48}=\frac{3}{144}$, and $\frac{25}{48}=\frac{75}{144}$.

(19.) L. C. M. of 7, 10, 12, 35, 63, and 28 is 1260: $\frac{1}{7}=\frac{180}{1260}$, and $\frac{4}{7}=\frac{720}{1260}$: $\frac{1}{10}=\frac{126}{1260}$, and $\frac{3}{10}=\frac{378}{1260}$: $\frac{1}{12}=\frac{105}{1260}$, and $\frac{5}{12}=\frac{525}{1260}$: $\frac{1}{35}=\frac{36}{1260}$, and $\frac{17}{35}=\frac{612}{1260}$: $\frac{1}{63}=\frac{20}{1260}$, and $\frac{4}{63}=\frac{80}{1260}$: $\frac{1}{28}=\frac{45}{1260}$, and $\frac{15}{28}=\frac{675}{1260}$.

(20.) L. C. M. of 5, 10, 25, 30, 45, and 60 is 900; $\frac{1}{5}=\frac{180}{900}$, and $\frac{3}{5}=\frac{540}{900}$: $\frac{1}{10}=\frac{90}{900}$, and $\frac{7}{10}=\frac{630}{900}$: $\frac{1}{25}=\frac{36}{900}$, and $\frac{6}{25}=\frac{216}{900}$: $\frac{1}{30}=\frac{30}{900}$, and $\frac{11}{30}=\frac{330}{900}$: $\frac{1}{45}=\frac{20}{900}$, and $\frac{13}{45}=\frac{260}{900}$: $\frac{1}{60}=\frac{15}{900}$, and $\frac{23}{60}=\frac{345}{900}$.

Art. 110.

(6.) $\frac{3}{11}+\frac{7}{11}+\frac{8}{11}+\frac{10}{11}=\frac{28}{11}=2\frac{6}{11}$, *Ans.*

(7.) $\frac{5}{13}+\frac{8}{13}+\frac{9}{13}+\frac{11}{13}=\frac{33}{13}=2\frac{7}{13}$, *Ans.*

(8.) $\frac{7}{15}+\frac{8}{15}+\frac{11}{15}+\frac{13}{15}=\frac{39}{15}=2\frac{9}{15}=2\frac{3}{5}$, *Ans.*

(9.) $\frac{9}{20}+\frac{11}{20}+\frac{13}{20}+\frac{17}{20}=\frac{50}{20}=2\frac{10}{20}=2\frac{1}{2}$, *Ans.*

(10.) $\frac{12}{25}+\frac{16}{25}+\frac{18}{25}+\frac{24}{25}=\frac{70}{25}=2\frac{20}{25}=2\frac{4}{5}$, *Ans.*

Art. 111.

(2.) The least common denominator is 6; $\frac{1}{2}=\frac{3}{6}$, $\frac{1}{3}=\frac{2}{6}$: $\frac{3}{6}+\frac{2}{6}=\frac{5}{6}$, *Ans.*

(4.) The L. C. D. is 10; $\frac{1}{2}=\frac{5}{10}$, $\frac{3}{5}=\frac{6}{10}$: $\frac{6}{10}+\frac{5}{10}=\frac{11}{10}=1\frac{1}{10}$, *Ans.*

(8.) $2\frac{1}{2}=\frac{5}{2}$, $3\frac{2}{3}=\frac{11}{3}$; the L. C. D. $=6$; $\frac{5}{2}=\frac{15}{6}$, $\frac{11}{3}=\frac{22}{6}$: $\frac{15}{6}+\frac{22}{6}=\frac{37}{6}=6\frac{1}{6}$, *Ans.*

(9.) L. C. D. $=12$; $\frac{2}{3}=\frac{8}{12}$, $\frac{3}{4}=\frac{9}{12}$, $\frac{5}{6}=\frac{10}{12}$: $\frac{8+9+10}{12}=\frac{27}{12}=2\frac{3}{12}=2\frac{1}{4}$, *Ans.*

(10.) L. C. D. $=24$; $\frac{1}{4}=\frac{6}{24}$, $\frac{7}{8}=\frac{21}{24}$, $\frac{11}{12}=\frac{22}{24}$: $\frac{6+21+22}{24}=\frac{49}{24}=2\frac{1}{24}$, *Ans.*

(11.) L. C. D. $= 792$; $\frac{1}{8} = \frac{99}{792}$, $\frac{1}{9} = \frac{88}{792}$, $\frac{2}{11} = \frac{144}{792}$: $\frac{99+88+144}{792} = \frac{331}{792}$, *Ans.*

(12.) $\frac{4}{5} = \frac{16}{20}$, $\frac{1}{2} = \frac{10}{20}$, $\frac{3}{4} = \frac{15}{20}$; $\frac{16}{20} + \frac{10}{20} + \frac{15}{20} = \frac{41}{20} = 2\frac{1}{20}$: $7 + 8 + 2\frac{1}{20} = 17\frac{1}{20}$, *Ans.*

(13.) L. C. D. $= 5460$; $\frac{1}{12} = \frac{455}{5460}$, $\frac{1}{13} = \frac{420}{5460}$, $\frac{1}{14} = \frac{390}{5460}$, $\frac{1}{15} = \frac{364}{5460}$: $\frac{455+420+390+364}{5460} = \frac{1629}{5460} = \frac{543}{1820}$, *Ans.*

(14.) L. C. D. $= 180$; $\frac{13}{18} = \frac{130}{180}$, $\frac{8}{15} = \frac{96}{180}$, $\frac{11}{20} = \frac{99}{180}$, $\frac{13}{30} = \frac{78}{180}$: $\frac{130+96+99+78}{180} = \frac{403}{180} = 2\frac{43}{180}$, *Ans.*

(15)

$\frac{7}{12}$	$\frac{7}{12} = \frac{42}{72}$
$2\frac{5}{6}$	$\frac{5}{6} = \frac{60}{72}$
$3\frac{3}{8}$	$\frac{3}{8} = \frac{27}{72}$
$3\frac{4}{9}$	$\frac{4}{9} = \frac{32}{72}$
8	$\frac{161}{72} = 2\frac{17}{72}$
$2\frac{17}{72}$	
$10\frac{17}{72}$, *Ans.*	

(16)

$16\frac{2}{3}$	$\frac{2}{3} = \frac{40}{60}$
$12\frac{3}{4}$	$\frac{3}{4} = \frac{45}{60}$
$8\frac{3}{5}$	$\frac{3}{5} = \frac{36}{60}$
$2\frac{1}{4}$	$\frac{1}{4} = \frac{15}{60}$
38	$\frac{136}{60} = 2\frac{4}{15}$
$2\frac{4}{15}$	
$40\frac{4}{15}$, *Ans.*	

(17.) L. C. D. $= 60$; $\frac{1}{2} = \frac{30}{60}$, $\frac{1}{3} = \frac{20}{60}$, $\frac{1}{4} = \frac{15}{60}$, $\frac{1}{5} = \frac{12}{60}$, $\frac{1}{6} = \frac{10}{60}$: $\frac{30+20+15+12+10}{60} = \frac{87}{60} = 1\frac{27}{60} = 1\frac{9}{20}$, *Ans.*

(18.) $\frac{2}{5} = \frac{1120}{2800}$, $\frac{7}{16} = \frac{1225}{2800}$, $\frac{7}{50} = \frac{392}{2800}$, $\frac{3}{140} = \frac{60}{2800}$, $\frac{3}{2800}$: $\frac{1120+1225+392+60+3}{2800} = \frac{2800}{2800} = 1$, *Ans.*

(19.) $\frac{1}{20} = \frac{36}{720}$, $\frac{7}{16} = \frac{315}{720}$, $\frac{11}{12} = \frac{660}{720}$, $\frac{2}{15} = \frac{96}{720}$, $\frac{11}{18} = \frac{440}{720}$: $\frac{36}{720} + \frac{315}{720} + \frac{660}{720} + \frac{96}{720} + \frac{440}{720} = \frac{1547}{720} = 2\frac{107}{720}$: $1 + 2 + 2\frac{107}{720} = 5\frac{107}{720}$, *Ans.*

(20.) $\frac{2}{3} = \frac{40}{60}$, $\frac{1}{2} = \frac{30}{60}$, $\frac{1}{5} = \frac{12}{60}$, $\frac{1}{3} = \frac{20}{60}$, $\frac{1}{4} = \frac{15}{60}$: $\frac{40}{60} + \frac{30}{60} + \frac{12}{60} + \frac{20}{60} + \frac{15}{60} = \frac{117}{60} = 1\frac{57}{60} = 1\frac{19}{20}$: $2 + 4 + 6 + 8 + 1\frac{19}{20} = 21\frac{19}{20}$, *Ans.*

(21.) $\frac{1}{3} = \frac{35}{105}$, $\frac{2}{7} = \frac{30}{105}$, $\frac{1}{5} = \frac{21}{105}$, $\frac{1}{21} = \frac{5}{105}$: $\frac{35}{105} + \frac{30}{105} + \frac{21}{105} + \frac{5}{105} = \frac{91}{105} = \frac{13}{15}$: $1 + 4 + 2 + 2 + \frac{13}{15} = 9\frac{13}{15}$, *Ans.*

Art. 113.

(2.) $\frac{3}{4}-\frac{1}{4}=\frac{2}{4}=\frac{1}{2}$, *Ans.*

(7.) $4\frac{1}{4}-2\frac{3}{4}$. $\frac{3}{4}$ can not be taken from $\frac{1}{4}$; so borrow 1 from 4. $1=\frac{4}{4}$; $\frac{4}{4}+\frac{1}{4}=\frac{5}{4}$; $\frac{3}{4}$ from $\frac{5}{4}=\frac{2}{4}$ or $\frac{1}{2}$. Since we took 1 from 4, only 3 remain, and $3-2=1$. *Ans.* $1\frac{1}{2}$.

(8)

$$\begin{array}{l} 8\frac{1}{3} \\ 3\frac{2}{3} \\ \hline 4\frac{2}{3}\text{, } Ans. \end{array} \qquad \begin{array}{l} \frac{1}{3}+\frac{3}{3}=\frac{4}{3} \\ \frac{4}{3}-\frac{2}{3}=\frac{2}{3} \end{array}$$

(9)

$$\begin{array}{l} 23\frac{7}{20} \\ 17\frac{11}{20} \\ \hline 5\frac{4}{5}\text{, } Ans. \end{array} \qquad \begin{array}{l} \frac{7}{20}+\frac{20}{20}=\frac{27}{20} \\ \frac{27}{20}-\frac{11}{20}=\frac{16}{20}=\frac{4}{5} \end{array}$$

Art. 114.

(9.) L. C. D. $=30$; $\frac{4}{15}=\frac{8}{30}$, $\frac{1}{10}=\frac{3}{30}$: $\frac{8}{30}-\frac{3}{30}=\frac{5}{30}=\frac{1}{6}$, *Ans.*

(10.) L. C. D. $=42$; $\frac{16}{21}=\frac{32}{42}$, $\frac{5}{14}=\frac{15}{42}$: $\frac{32}{42}-\frac{15}{42}=\frac{17}{42}$, *Ans.*

(12.) $5=\frac{15}{3}$: $\frac{15}{3}-\frac{2}{3}=\frac{13}{3}=4\frac{1}{3}$, *Ans.*

(13.) $5\frac{2}{3}=\frac{17}{3}=\frac{34}{6}$, $4\frac{1}{2}=\frac{9}{2}=\frac{27}{6}$: $\frac{34}{6}-\frac{27}{6}=\frac{7}{6}=1\frac{1}{6}$, *Ans.*

(14.) $7\frac{2}{3}=\frac{23}{3}=\frac{92}{12}$, $4\frac{3}{4}=\frac{19}{4}=\frac{57}{12}$: $\frac{92}{12}-\frac{57}{12}=\frac{35}{12}=2\frac{11}{12}$, *Ans.*

(15.) $14\frac{1}{4}=\frac{57}{4}=\frac{171}{12}$, $12\frac{2}{3}=\frac{38}{3}=\frac{152}{12}$: $\frac{171}{12}-\frac{152}{12}=\frac{19}{12}=1\frac{7}{12}$, *Ans.*

(16.) $5\frac{3}{14}=\frac{73}{14}=\frac{219}{42}$, $2\frac{10}{21}=\frac{52}{21}=\frac{104}{42}$: $\frac{219}{42}-\frac{104}{42}=\frac{115}{42}=2\frac{31}{42}$, *Ans.*

(17.) $4\frac{1}{24}=\frac{97}{24}=\frac{194}{48}$, $3\frac{1}{16}=\frac{49}{16}=\frac{147}{48}$: $\frac{194}{48}-\frac{147}{48}=\frac{47}{48}$, *Ans.*

(18.) $56\frac{1}{3}=\frac{169}{3}=\frac{676}{12}$, $42\frac{1}{4}=\frac{169}{4}=\frac{507}{12}$: $\frac{676}{12}-\frac{507}{12}=\frac{169}{12}=14\frac{1}{12}$, *Ans.*

(19.) $60\frac{4}{5}=\frac{304}{5}=\frac{608}{10}$, $41\frac{3}{10}=\frac{413}{10}$: $\frac{608}{10}-\frac{413}{10}=\frac{195}{10}=19\frac{1}{2}$, *Ans.*

(20.) $97\frac{1}{2}=\frac{195}{2}=\frac{585}{6}$, $48\frac{5}{6}=\frac{293}{6}$: $\frac{585}{6}-\frac{293}{6}=\frac{292}{6}=48\frac{2}{3}$, *Ans.*

Art. 115.

(5.) $\frac{3}{4} \times 3 = \frac{9}{4} = 2\frac{1}{4}$, *Ans.*

(6.) $8 \times \frac{2}{3} = \frac{16}{3} = 5\frac{1}{3}$, *Ans.*

(7.) $\frac{3}{4} \times \frac{5}{7} = \frac{15}{28}$, *Ans.*

(8.) $\frac{2}{3} \times 4 = \frac{8}{3} = 2\frac{2}{3}$, *Ans.*

(9.) $5 \times \frac{3}{4} = \frac{15}{4} = 3\frac{3}{4}$, *Ans.*

(11.) $\frac{2}{3} \times 6 = \frac{12}{3} = 4$, *Ans.*

(12.) $20 \times \frac{3}{4} = \frac{60}{4} = 15$, *Ans.*

(13.) $\frac{\cancel{8}}{13} \times \frac{11}{\underset{2}{\cancel{16}}} = \frac{11}{26}$, *Ans.*

(14.) $\frac{3}{5} \times 10 = \frac{30}{5} = 6$, *Ans.*

(15.) $12 \times \frac{2}{3} = \frac{24}{3} = 8$, *Ans.*

(16.) $\frac{9}{13} \times \frac{3}{7}$: $\frac{9}{13} \times \frac{3}{7} = \frac{27}{91}$, *Ans.*

(17.) $\frac{3}{7} \times 6 = \frac{18}{7} = 2\frac{4}{7}$, *Ans.*

(18.) $7 \times \frac{2}{3} = \frac{14}{3} = 4\frac{2}{3}$, *Ans.*

(21.) 8 times $3 = 24$: 8 times $\frac{2}{3} = \frac{16}{3} = 5\frac{1}{3}$: $24 + 5\frac{1}{3} = 29\frac{1}{3}$, *Ans.*

(22.) $2\frac{1}{2} = \frac{5}{2}$: $\frac{5}{2} \times \frac{5}{2} = \frac{25}{4} = 6\frac{1}{4}$, *Ans.*

(23.) $10 \times 7 = 70$: $\frac{7}{9} \times 7 = \frac{49}{9} = 5\frac{4}{9}$: $70 + 5\frac{4}{9} = 75\frac{4}{9}$, *Ans.*

(24.) $25 \times 8 = 200$: $25 \times \frac{3}{5} = \frac{75}{5} = 15$: $200 + 15 = 215$, *Ans.*

(25.)

$$17\frac{3}{11} = \frac{190}{11} : \frac{9}{\cancel{10}} \times \frac{\overset{19}{\cancel{190}}}{11} = \frac{171}{11} = 15\frac{6}{11}, \textit{ Ans.}$$

(26.) $10 \times 9 = 90$: $\frac{5}{6} \times 9 = \frac{45}{6} = 7\frac{3}{6} = 7\frac{1}{2}$: $90 + 7\frac{1}{2} = 97\frac{1}{2}$, *Ans.*

(27.) 8 times $64=512$: $\frac{1}{8}$ of $64=8$: $\frac{7}{8}=56$: $512+56=568$, *Ans.*

(28.) $8\frac{3}{4}=\frac{35}{4}$: $\frac{1}{7}$ of $\frac{35}{4}=\frac{5}{4}$: $\frac{3}{7}=\frac{15}{4}=3\frac{3}{4}$, *Ans.*

(29.)

$$2\frac{2}{11}=\frac{24}{11}:\quad \frac{5}{\underset{4}{\cancel{12}}}\times\frac{9}{\underset{2}{\cancel{16}}}\times\frac{\overset{3}{\cancel{24}}}{11}=\frac{45}{88},\ \textit{Ans.}$$

(30.)

$$2\frac{1}{16}=\frac{33}{16}:\quad \frac{\overset{3}{\cancel{33}}}{\cancel{16}}\times\frac{\cancel{3}}{\cancel{11}}\times\frac{\cancel{16}}{\underset{3}{\cancel{9}}}=1,\ \textit{Ans.}$$

(31.)

$$\frac{\overset{3}{\cancel{27}}}{\underset{2}{\cancel{4}}}\times\frac{\overset{13}{\cancel{26}}}{\cancel{9}}\times\frac{21}{1}=\frac{819}{2}=409\frac{1}{2},\ \textit{Ans.}$$

(32.)

$$\frac{5}{\cancel{2}}\times\frac{11}{3}\times\frac{19}{\cancel{4}}\times\frac{\overset{2}{\cancel{8}}}{7}=\frac{1045}{21}=49\frac{16}{21},\ \textit{Ans.}$$

(33.)

$$\frac{\cancel{11}}{\cancel{5}}\times\frac{\overset{11}{\cancel{55}}}{\underset{2}{\cancel{26}}}\times\frac{\cancel{13}}{\cancel{4}}\times\frac{\overset{\overset{2}{\cancel{4}}}{\cancel{16}}}{\cancel{11}}=22,\ \textit{Ans.}$$

(34) $$\frac{\cancel{7}}{\cancel{8}}\times\frac{\cancel{3}}{\underset{2}{\cancel{10}}}\times\frac{\cancel{8}}{9}\times\frac{\cancel{5}}{\cancel{6}}\times\frac{\cancel{2}}{\cancel{3}}\times\frac{\cancel{6}}{\cancel{7}}=\frac{1}{9},\ \textit{Ans.}$$

(35.)

$$\frac{\cancel{1}}{\underset{2}{\cancel{4}}}\times\frac{\cancel{9}}{\cancel{7}}\times\frac{\cancel{4}}{\cancel{5}}\times\frac{\cancel{7}}{\cancel{9}}\times\frac{\cancel{5}}{\cancel{4}}\times\frac{\cancel{2}}{\cancel{3}}\times\frac{\overset{2}{\cancel{6}}}{\cancel{1}}=1,\ \textit{Ans.}$$

(36.)

$$\frac{\cancel{6}}{\cancel{7}}\times\frac{\cancel{4}}{9}\times\frac{\cancel{7}}{\cancel{4}}\times\frac{\cancel{1}}{\underset{\cancel{3}}{\cancel{6}}}\times\frac{\cancel{3}}{\cancel{4}}\times\frac{\cancel{5}}{\cancel{6}}\times\frac{\cancel{2}}{\cancel{5}}\times\frac{\overset{5}{\cancel{20}}}{\cancel{1}}=\tfrac{5}{9}, \textit{ Ans.}$$

(37.)

$$\frac{\cancel{5}}{\cancel{2}}\times\frac{\overset{8}{\cancel{32}}}{\cancel{5}}\times\frac{\cancel{13}}{\cancel{4}}\times\frac{\cancel{7}}{\cancel{13}}\times\frac{\cancel{2}}{1}\times\frac{3}{\cancel{7}}=24, \textit{ Ans.}$$

Art. 116.

(2.) $\frac{1}{4}$ of $5 = \frac{5}{4}$; then $\frac{3}{4}$ of $5 = 3$ times $\frac{5}{4} = \frac{15}{4} = 3\frac{3}{4}$, *Ans.*

(3.) $\frac{2}{5}$ of $7 = \frac{14}{5} = 2\frac{4}{5}$, *Ans.*

(4.) $\frac{4}{5}$ of $10 = \frac{40}{5} = 8$, *Ans.*

(5.) $\frac{1}{6}$ of $12 = 2$: $\frac{5}{6} = 2 \times 5 = 10$, *Ans.*

(6.) $\frac{5}{6}$ of $15 = \frac{75}{6} = 12\frac{3}{6} = 12\frac{1}{2}$, *Ans.*

(7.) $\frac{8}{9}$ of $21 = \frac{168}{9} = 18\frac{6}{9} = 18\frac{2}{3}$, *Ans.*

(8.) $\frac{1}{10}$ of $25 = \frac{25}{10} = \frac{5}{2}$: $\frac{7}{10} = \frac{35}{2} = 17\frac{1}{2}$, *Ans.*

(9.) $\frac{5}{12}$ of $27 = \frac{135}{12} = 11\frac{3}{12} = 11\frac{1}{4}$, *Ans.*

(10.) $\frac{7}{12}$ of $28 = \frac{196}{12} = 16\frac{4}{12} = 16\frac{1}{3}$, *Ans.*

Art. 117.

(4.) $\frac{1}{2}$ of $\frac{3}{5}$ of $\frac{11}{4} = \frac{1}{2} \times \frac{3}{5} \times \frac{11}{4} = \frac{33}{40}$, *Ans.*

(7.) $\frac{2}{3}$ of $\frac{5}{7}$ of $\frac{13}{9} = \frac{2}{3} \times \frac{5}{7} \times \frac{13}{9} = \frac{130}{189}$, *Ans.*

(8.) $\frac{2}{\cancel{3}}$ of $\frac{\cancel{3}}{\cancel{4}}$ of $\frac{\cancel{4}}{5} = \frac{2}{5}$, *Ans.*

(9.) $\frac{1}{\cancel{3}}$ of $\frac{\cancel{3}}{4}$ of $\frac{5}{6} = \frac{5}{24}$, *Ans.*

(10.) $\frac{3}{\cancel{5}}$ of $\frac{\cancel{5}}{\cancel{7}}$ of $\frac{\cancel{7}}{8} = \frac{3}{8}$, *Ans.*

(11.)

$$\frac{\cancel{3}}{5} \text{ of } \frac{\cancel{4}}{\cancel{9}} \text{ of } \frac{\cancel{7}}{\underset{3}{\cancel{12}}} \text{ of } \frac{\overset{2}{\cancel{18}}}{\underset{5}{\cancel{35}}} = \tfrac{2}{25}, \textit{ Ans.}$$

(12.) $\frac{1}{\cancel{3}} \text{ of } \frac{\cancel{3}}{\cancel{4}} \text{ of } \frac{\cancel{4}}{9} = \tfrac{1}{9}$, *Ans.*

(13.) $\frac{1}{9} \text{ of } \frac{\cancel{3}}{\cancel{4}} \text{ of } \frac{\cancel{4}}{\cancel{3}} = \tfrac{1}{9}$, *Ans.*

(14.)

$$\frac{\cancel{3}}{\cancel{5}} \text{ of } \frac{\cancel{6}}{\cancel{7}} \text{ of } \frac{\overset{\cancel{5}}{\cancel{35}}}{\underset{\cancel{3}}{\cancel{18}}} = 1, \textit{ Ans.}$$

(15.)

$$\frac{\cancel{3}}{\cancel{7}} \text{ of } \frac{\overset{2}{\cancel{8}}}{\cancel{3}} \text{ of } \frac{\cancel{7}}{\cancel{4}} = 2, \textit{ Ans.}$$

(16.) $\frac{\cancel{9}}{\cancel{13}} \text{ of } \frac{\cancel{7}}{\underset{2}{\cancel{18}}} \text{ of } \frac{\cancel{13}}{\cancel{7}} = \tfrac{1}{2}$, *Ans.*

(17.) $\frac{1}{2} \text{ of } \frac{\cancel{4}}{\cancel{5}} \text{ of } \frac{1}{\underset{2}{\cancel{8}}} \text{ of } \frac{\cancel{5}}{1} = \tfrac{1}{4}$, *Ans.*

(18.) $\frac{1}{\cancel{2}} \text{ of } \frac{\cancel{2}}{\cancel{3}} \text{ of } \frac{\cancel{3}}{\cancel{4}} \text{ of } \frac{\cancel{4}}{\cancel{5}} \text{ of } \frac{\cancel{5}}{8} \text{ of } \frac{\cancel{5}}{\cancel{9}} \text{ of } \frac{\cancel{9}}{\underset{2}{\cancel{10}}} = \tfrac{1}{16}$, *Ans.*

Art. 118.

(1.)

$$2\tfrac{1}{3} = \tfrac{7}{3},\ 13\tfrac{1}{5} = \tfrac{66}{5}:\ \frac{7}{\cancel{3}} \times \frac{\overset{22}{\cancel{66}}}{5} = \tfrac{154}{5} = 30\tfrac{4}{5} \text{ ct., } \textit{Ans.}$$

(2.) 3 times $\frac{2}{3} = \frac{6}{3} = \$2$: 5 times $\frac{2}{3} = \frac{10}{3} = \$3\frac{1}{3}$: 7 times $\frac{2}{3} = \frac{14}{3} = \$4\frac{2}{3}$: $\frac{13}{\cancel{2}} \times \frac{\cancel{2}}{3} = \frac{13}{3} = \$4\frac{1}{3}$: $\frac{23}{\underset{2}{\cancel{4}}} \times \frac{\cancel{2}}{3} = \frac{23}{6}$ $= \$3\frac{5}{6}$.

(3.) $\frac{\overset{2}{\cancel{10}}}{\cancel{3}} \times \frac{\overset{8}{\cancel{24}}}{\cancel{5}} = 16$ ct., *Ans.*

(4.) $\frac{\overset{4}{\cancel{16}}}{\cancel{5}} \times \frac{\overset{15}{\cancel{75}}}{\cancel{4}} = \$60$, *Ans.*

(5.) $\frac{\cancel{5}}{\cancel{3}} \times \frac{\cancel{3}}{\underset{4}{\cancel{20}}} = \$\frac{1}{4}$, *Ans.*

(6.) $\frac{\cancel{5}}{\cancel{2}} \times \frac{\overset{2}{\cancel{4}}}{\cancel{5}} = \$2$, *Ans.*

(7.) $\frac{\overset{10}{\cancel{50}}}{\underset{3}{\cancel{9}}} \times \frac{\overset{2}{\cancel{6}}}{\cancel{5}} = \frac{20}{3} = \$6\frac{2}{3}$, *Ans.*

(8.) $\frac{11}{2} \times \frac{31}{4} = \frac{341}{8} = 42\frac{5}{8}$ mi., *Ans.*

(9.) $\frac{\cancel{3}}{5}$ of $\frac{2}{\cancel{3}} = \frac{2}{5}$, *Ans.*

(10.) $\frac{2}{9}$ of $\frac{11}{2} = \frac{11}{9}$: $\frac{11}{\cancel{9}} \times \frac{\overset{3}{\cancel{27}}}{4} = \frac{33}{4} = \$8\frac{1}{4}$, *Ans.*

(11.) $\frac{\cancel{3}}{\cancel{7}} \times \frac{5}{\underset{\cancel{3}}{\cancel{9}}} \times \frac{\overset{11}{\cancel{33}}}{\cancel{2}} \times \frac{\cancel{2}}{\cancel{3}} \times \frac{\cancel{7}}{8} \times \frac{\overset{5}{\cancel{15}}}{1} = \frac{275}{8} = 34\frac{3}{8}$, *Ans.*

(12.) $\frac{2}{3} = \frac{8}{12}$, $\frac{3}{4} = \frac{9}{12}$: $\frac{8}{12} + \frac{9}{12} = \frac{17}{12}$: $\frac{\cancel{2}}{\cancel{3}} \times \frac{\cancel{3}}{\underset{2}{\cancel{4}}} = \frac{1}{2}$: $\frac{1}{2} = \frac{6}{12}$: $\frac{17}{12} \times \frac{6}{12} = \frac{23}{12} = 1\frac{11}{12}$. *Ans.*

Art. 119.

REMARK.—Pupils are often at a loss to understand, why it is that the quotient of one proper fraction, divided by another, is sometimes a whole number, or greater than unity. The teacher should be careful to explain this subject, by means of familiar examples, such as may be found in "Ray's New Intellectual Arithmetic," Lessons XXXIII—XXXVIII.

It should also be shown, that if we take any dividend, and divide it by different numbers, that as the divisor becomes less, the quotient becomes greater; so that, by making the divisor sufficiently small, the quotient may be made as large as we please. Thus, the quotient of $\frac{1}{2}$ divided by $\frac{1}{4}$ is 2; by $\frac{1}{8}$, is 4; by $\frac{1}{16}$ is 8; by $\frac{1}{4000000}$, is 2000000, etc. It is on this principle, that mathematicians, say, that the quotient of any number, divided by 0, is infinitely large.

(6.) 1 yd. will cost $\frac{1}{4}$ of $\$\frac{8}{9} = \$\frac{2}{9}$, *Ans.*

(7.) $3 \div \frac{1}{2} = 3 \times \frac{2}{1} = 6$, *Ans.*

(8.) $\frac{9}{10} \div \frac{1}{5} = \frac{9}{\underset{2}{\cancel{10}}} \times \frac{\cancel{5}}{1} = \frac{9}{2} = 4\frac{1}{2}$ yd., *Ans.*

(9.) One cent will buy $\frac{1}{3}$ of an orange: $\frac{1}{2}$ cent will buy $\frac{1}{2}$ of $\frac{1}{3} = \frac{1}{6}$, *Ans.*

(10.) $6 \div \frac{3}{4} = 6 \times \frac{4}{3} = \frac{24}{3} = 8$ yd., *Ans.*

(11.) $\frac{3}{4} \div \frac{1}{5} = \frac{3}{4} \times \frac{5}{1} = \frac{15}{4} = 3\frac{3}{4}$ yd., *Ans.*

(12.) 1 lb. will cost $\frac{1}{7}$ of $\$\frac{14}{25} = \$\frac{2}{25}$, *Ans.*

(14.)

$2\frac{2}{5} = \frac{12}{5}$: $\frac{1}{\cancel{6}}$ of $\frac{\overset{2}{\cancel{12}}}{5} = \frac{2}{5}$, *Ans.*

(15.)

$5\frac{1}{2} = \frac{11}{2}$: $\frac{\overset{2}{\cancel{22}}}{1} \times \frac{2}{\cancel{11}} = 4$, *Ans.*

(16.) $\frac{5}{\cancel{2}} \times \frac{\overset{8}{\cancel{16}}}{1} = 40$, *Ans.*

(17.) $\frac{\overset{3}{\cancel{24}}}{5} \times \frac{1}{\cancel{8}} = \frac{3}{5}$, *Ans.*

(18.) $\frac{\cancel{6}}{1} \times \frac{5}{\underset{2}{\cancel{12}}} = \frac{5}{2} = 2\frac{1}{2}$, *Ans.*

(19.) $\frac{19}{\cancel{4}} \times \frac{\overset{2}{\cancel{8}}}{41} = \frac{38}{41}$, *Ans.*

(20.) $\frac{\overset{8}{\cancel{88}}}{7} \times \frac{1}{\cancel{11}} = \frac{8}{7} = 1\frac{1}{7}$, *Ans.*

(21.) $\frac{\overset{2}{\cancel{30}}}{1} \times \frac{4}{\cancel{15}} = 8$, *Ans.*

(22.) $\frac{\overset{3}{\cancel{9}}}{\underset{2}{\cancel{4}}} \times \frac{\cancel{2}}{\underset{5}{\cancel{15}}} = \frac{3}{10}$, *Ans.*

(23.) $\frac{11}{3} \times \frac{1}{7} = \frac{11}{21}$, *Ans.*

(24.) $\frac{50}{1} \times \frac{7}{31} = \frac{350}{31} = 11\frac{9}{31}$, *Ans.*

(25.) $\frac{1}{2} \times \frac{\overset{25}{\cancel{50}}}{1} = 25$, *Ans.*

(26.) $\frac{237}{5} \times \frac{1}{15} = \frac{237}{75} = 3\frac{12}{75} = 3\frac{4}{25}$, *Ans.*

(27.) $\frac{\overset{8}{\cancel{56}}}{1} \times \frac{9}{\underset{7}{\cancel{49}}} = \frac{72}{7} = 10\frac{2}{7}$, *Ans.*

(28.) $\frac{\overset{2}{\cancel{14}}}{15} \times \frac{1}{\underset{3}{\cancel{21}}} = \frac{2}{45}$, *Ans.*

(29.) $\frac{392}{3} \times \frac{1}{18} = \frac{392}{54} = \frac{196}{27} = 7\frac{7}{27}$, *Ans.*

(31.) $$\frac{\cancel{3}}{5} \times \frac{\overset{4}{\cancel{8}}}{\underset{3}{\cancel{9}}} \times \frac{7}{\underset{3}{\cancel{6}}} \times \frac{4}{3} = \frac{112}{135}, \textit{ Ans.}$$

(32.) $$\frac{1}{3} \times \frac{41}{\underset{2}{\cancel{8}}} \times \frac{\cancel{4}}{3} \times \frac{\cancel{2}}{35} = \frac{41}{315}, \textit{ Ans.}$$

(33.) $$\frac{\cancel{5}}{\underset{6}{\cancel{18}}} \times \frac{\cancel{2}}{\cancel{5}} \times \frac{\overset{3}{\cancel{123}}}{\underset{2}{\cancel{10}}} \times \frac{\cancel{5}}{1} \times \frac{5}{\cancel{41}} = \frac{5}{6}, \textit{ Ans.}$$

(34.) $$\frac{\cancel{2}}{\cancel{7}} \times \frac{\cancel{7}}{\underset{2}{\cancel{8}}} \times \frac{\cancel{4}}{\cancel{3}} \times \frac{\cancel{3}}{1} \times \frac{1}{5} = \frac{1}{5}, \textit{ Ans.}$$

(35.) $$\frac{\cancel{5}}{\underset{6}{\cancel{18}}} \times \frac{\cancel{2}}{\cancel{5}} \times \frac{\overset{3}{\cancel{123}}}{\cancel{10}} \times \frac{\cancel{5}}{1} \times \frac{\cancel{10}}{\cancel{41}} \times \frac{1}{\underset{\underset{2}{\cancel{4}}}{\cancel{20}}} = \frac{1}{12}, \textit{ Ans.}$$

Art. 120.

(6.) $\frac{3}{4} \times \frac{1}{5} = \frac{3}{20}$, *Ans.*

(7.) $$\frac{1}{\underset{2}{\cancel{4}}} \times \frac{\cancel{2}}{1} = \frac{1}{2}, \textit{ Ans.}$$

(8.) $$\frac{2}{\cancel{3}} \times \frac{\overset{2}{\cancel{6}}}{5} = \frac{4}{5}, \textit{ Ans.}$$

(9.) $$\frac{\overset{3}{\cancel{15}}}{4} \times \frac{1}{\cancel{5}} = \frac{3}{4}, \textit{ Ans.}$$

(10.) $$\frac{5}{\underset{2}{\cancel{6}}} \times \frac{\overset{3}{\cancel{9}}}{8} = \frac{15}{16}, \textit{ Ans.}$$

(11.)

$$\frac{\overset{7}{\cancel{77}}}{9} \times \frac{1}{\cancel{11}} = \frac{7}{9}, \textit{ Ans.}$$

(12.)

$$\frac{\overset{3}{\cancel{21}}}{\underset{\underset{2}{\cancel{4}}}{\cancel{32}}} \times \frac{\overset{\overset{3}{\cancel{6}}}{\cancel{48}}}{\underset{5}{\cancel{35}}} = \frac{9}{10}, \textit{ Ans.}$$

Art. 121.

(2.) $\frac{6}{7} \times \frac{5}{11} = \frac{30}{77}$, *Ans.*

(3.) $\frac{2}{3} \times \frac{1}{5} = \frac{2}{15}$, *Ans.*

(4.) $\frac{2}{1} \times \frac{3}{11} = \frac{6}{11}$, *Ans.*

(5.) $\frac{25}{8} \times \frac{7}{33} = \frac{175}{264}$, *Ans.*

(6.) $\frac{7}{3} \times \frac{2}{9} = \frac{14}{27}$, *Ans.*

(7.)

$$\frac{\cancel{15}}{\cancel{4}} \times \frac{\overset{2}{\cancel{8}}}{\underset{3}{\cancel{45}}} = \frac{2}{3}, \textit{ Ans.}$$

(8.)

$$\frac{\overset{8}{\cancel{88}}}{\cancel{9}} \times \frac{\overset{3}{\cancel{27}}}{\underset{5}{\cancel{55}}} = \frac{24}{5} = 4\tfrac{4}{5}, \textit{ Ans.}$$

(9.)

$$\frac{\overset{7}{\cancel{35}}}{\cancel{4}} \times \frac{\overset{2}{\cancel{8}}}{\underset{9}{\cancel{45}}} = 1\tfrac{5}{9}, \textit{ Ans.}$$

(10.) $\frac{47}{6} \times \frac{11}{97} = \frac{517}{582}$, *Ans.*

Art. 122.

(1.) $3\frac{1}{4} \div \frac{1}{2} = \frac{13}{4} \times \frac{2}{1} = \frac{13}{2} = 6\frac{1}{2}$ yd., *Ans.*

(2.) $2\frac{3}{10} \div \frac{3}{5} = \frac{23}{10} \times \frac{5}{3} = \frac{23}{6} = 3\frac{5}{6}$ lb., *Ans.*

(3.) $42\frac{1}{2} \div 3\frac{3}{4} = \frac{85}{2} \div \frac{15}{4} = \frac{85}{2} \times \frac{4}{15} = \frac{17}{1} \times \frac{2}{3} = \frac{34}{3} = 11\frac{1}{3}$ yd., *Ans.*

(4.) $10 \div \frac{3}{8} = \frac{10}{1} \times \frac{8}{3} = \frac{80}{3} = 26\frac{2}{3}$, *Ans.*

(5.) $\frac{3}{7}$ of $1\frac{1}{2} = \frac{3}{7}$ of $\frac{3}{2} = \frac{9}{14}$: $3\frac{3}{7} = \frac{24}{7}$: $\frac{24}{7} \div \frac{9}{14} = \frac{24}{7} \times \frac{14}{9} = \frac{8}{1} \times \frac{2}{3} = \frac{16}{3} = 5\frac{1}{3}$, *Ans.*

(6.) $\frac{4}{11}$ of $27\frac{1}{2} = \frac{4}{11}$ of $\frac{55}{2} = 10$: $\frac{3}{10}$ of $21\frac{1}{4} = \frac{3}{10}$ of $\frac{85}{4} = \frac{51}{8}$: $10 \div \frac{51}{8} = \frac{10}{1} \times \frac{8}{51} = \frac{80}{51} = 1\frac{29}{51}$, *Ans.*

(7.) $\frac{3}{\not 2} \times \frac{\not 3}{7} \times \frac{\not 3}{7} \times \frac{\not 2}{\underset{\not 3}{\not 9}} = \frac{3}{49}$, *Ans.*

(8.) $\frac{\not{113}}{\not{15}} \times \frac{\overset{4}{\not{12}}}{\not{113}} \times \frac{19}{\underset{\not 3}{\not 9}} \times \frac{\not{15}}{47} \times \frac{\not 5}{\not 4} \times \frac{\overset{2}{\not 6}}{\not 5} = \frac{38}{47}$, *Ans.*

(9.) $\frac{1\frac{1}{2}}{\frac{2}{3}} = \frac{3}{2} \times \frac{3}{2} = \frac{9}{4}$: $\frac{2\frac{2}{5}}{2\frac{1}{6}} = \frac{12}{5} \times \frac{6}{13} = \frac{72}{65}$: $\frac{9}{4} \div \frac{72}{65} = \frac{9}{4} \times \frac{65}{72} = \frac{1}{4} \times \frac{65}{8} = \frac{65}{32} = 2\frac{1}{32}$, *Ans.*

(10.) $\frac{\not 5}{3} \times \frac{2}{\not 5} = \frac{2}{3}$: $\frac{\overset{2}{\not{36}}}{\not 7} \times \frac{\not 7}{\underset{33}{\not{594}}} = \frac{2}{33}$: $\frac{\not 2}{\not 3} \times \frac{\overset{11}{\not{33}}}{\not 2} = 11$, *Ans.*

Art. 123.

(1)	(2)	(3)	(4)	(5)
$\$16\frac{1}{16}$	$\$9\frac{1}{8}$	$\$50\frac{1}{4}$	$\$32.31\frac{1}{4}$	$\$5.81\frac{1}{4}$
$9\frac{1}{8}$	$4\frac{7}{16}$	$27\frac{3}{16}$	$15.12\frac{1}{2}$	$1.18\frac{3}{4}$
$5\frac{7}{16}$	$0\frac{3}{8}$	$\$23\frac{1}{16}$	$\$17.18\frac{3}{4}$	$\$4.62\frac{1}{2}$
$2\frac{13}{16}$	$1\frac{5}{8}$			
$\$33\frac{7}{16}$	$\$15\frac{9}{16}$			

(6.) $12\frac{1}{2} \times 9 = 108 + 4\frac{1}{2} = 112\frac{1}{2}$ ct. $= \$1.12\frac{1}{2}$, *Ans.*

(7.) $21 \times 6\frac{1}{4} = 126 + 5\frac{1}{4} = 131\frac{1}{4}$ ct. $= \$1.31\frac{1}{4}$, *Ans.*

(8.) $\$3.18\frac{3}{4} \times 15 = \$47.70 + \$0.11\frac{1}{4} = \$47.81\frac{1}{4}$, *Ans.*

(9.) $62\frac{1}{2} \times 5\frac{1}{2} = \frac{125}{2} \times \frac{11}{2} = \frac{1375}{4} = 343\frac{3}{4}$ ct. $= \$3.43\frac{3}{4}$, *Ans.*

(10.) $18\frac{3}{4} = \frac{75}{4}$: $12\frac{1}{2} = \frac{25}{2}$: $\frac{75}{4} \times \frac{25}{2} = \frac{1875}{8} = 234\frac{3}{8}$ ct. $= \$2.34\frac{3}{8}$, *Ans.*

(11.) $16\frac{2}{3} = \frac{50}{3}$: $13\frac{1}{2} = \frac{27}{2}$: $\frac{50}{3} \times \frac{27}{2} = 25 \times 9 = 225$ ct. $= \$2.25$, *Ans.*

(12.) $\$3.37\frac{1}{2} \times 10\frac{1}{4} = \frac{675}{2} \times \frac{41}{4} = \frac{27675}{8} = 3459\frac{3}{8}$ ct. $= \$34.59\frac{3}{8}$, *Ans.*

(13.) $17\frac{2}{3} = \frac{53}{3}$: $\frac{375}{1} \times \frac{53}{3} = 125 \times 53 = 6625$ ct. $= \$66.25$, *Ans.*

(14.) $225 \div 18\frac{3}{4} = \frac{225}{1} \times \frac{4}{75} = \frac{3}{1} \times \frac{4}{1} = 12$ yd., *Ans.*

(15.) $581\frac{1}{4} \div 37\frac{1}{2} = \frac{2325}{4} \times \frac{2}{75} = \frac{31}{2} \times \frac{1}{1} = 15\frac{1}{2}$ bu., *Ans.*

(16.) $\$11.56\frac{1}{4} \div 5 = \$2.31\frac{1}{4}$, *Ans.*

(17.) $\$31.06\frac{1}{4} \div 7 = \$4.43\frac{3}{4}$, *Ans.*

(18.) 5 mi. $\times$ 320 $=$ 1600 rd.: 1600 rd. $\times 16\frac{1}{2} = 26400$ ft.: 26400 ft. $\times$ 12 $=$316800 in., *Ans.*

(19.) 2 mi. $\times$ 320 $+$ 2 rd. $=$ 642 rd.: 642 rd. $\times 16\frac{1}{2} +$ 2 ft. $=$ 10595 ft., *Ans.*

(21.) 15875 ft. $\div 16\frac{1}{2} =$ 962 rd. 2 ft.: 962 rd. $\div$ 320 $=$ 3 mi. 2 rd. *Ans.* 3 mi. 2 rd. 2 ft.

(22.) 142634 in. $\div$ 12 $=$ 11886 ft. 2 in.: 11886 ft. $\div$ 3 $=$ 3962 yd.: 3962 yd. $\div 5\frac{1}{2} =$ 720 rd. 2 yd.: 720 rd. $\div$ 320 $=$ 2 mi. 80 rd. *Ans.* 2 mi. 80 rd. 2 yd. 2 in.

(23.) 2 mi. $=$ 126720 in.: 2 ft. 8 in. $=$ 32 in.: 126720 in. $\div$ 32 in. $=$ 3960, *Ans.*

(24.) 65 mi. $=$ 4118400 in.: 9 ft. 2 in. $=$ 110 in.: 4118400 in. $\div$ 110 in. $=$ 37440, *Ans.*

(25.) 1 A. $\times$ 160 $+$ 136 sq. rd. $=$ 296 sq. rd.: 296 sq. rd. $\times 30\frac{1}{4} +$ 25 sq. yd. $=$ 8979 sq. yd., *Ans.*

(26.) 7506 sq. yd. $\div 30\frac{1}{4} =$ 248 sq. rd. 4 sq. yd.: 248 sq. rd. $\div$ 160 $=$ 1 A. 88 sq. rd. *Ans.* 1 A. 88 sq. rd. 4 sq. yd.

(27.) 5 ch. 15 l. $=$ 515 l.: $7\frac{92}{100}$ in. $= \frac{792}{100}$ in.: $\frac{792}{100}$ in. $\times$ 515 $= \frac{407880}{100} = 4078\frac{4}{5}$ in., *Ans.*

(28.) $40\frac{1}{2} = \frac{81}{2}$: $\frac{81}{2} \times \frac{32}{1} = 81 \times 16 = 1296$ sq. rd.: 1296 sq. rd. $\div$ 160 $=$ 8 A. 16 sq. rd., *Ans.*

(29.) $365\frac{1}{4}$ da. $\times 4 = 1461$ da.: 1461 da. $\times 24 = 35064$ hr., *Ans.*

(30.) 914092 hr. $\div 24 = 38087$ da. 4 hr.: 38087 da. $\div 365\frac{1}{4} = 104$ yr. 101 da.: 104 yr. $\div 100 = 1$ cen. 4 yr. *Ans.* 1 cen. 4 yr. 101 da. 4 hr.

(31.) $238545 \div 31 = 7695$ da.: $7695 \div 365\frac{1}{4} = 21$ yr., and 99 quarter days remaining, which, reduced to days, by dividing by 4, makes $24\frac{3}{4}$ days. *Ans.* 21 yr. $24\frac{3}{4}$ da.

Art. 124.

(3.) $\frac{1}{28}$ lb. $\times 16 = \frac{16}{28} = \frac{4}{7}$ oz., *Ans.*

(4.) $\frac{1}{16}$ lb. $\times 12 = \frac{12}{16} = \frac{3}{4}$ oz., *Ans.*

(5.) $\frac{1}{20}$ rd. $\times 5\frac{1}{2} = \frac{11}{40}$ yd. $\times 3 = \frac{33}{40}$ ft., *Ans.*

(6.) $\frac{7}{1280}$ A. $\times 160 = \frac{7}{8}$ sq. rd., *Ans.*

(7.) $\$\frac{3}{350} \times 100 = \frac{300}{350} = \frac{6}{7}$ ct., *Ans.*

(8.) $\frac{1}{1584}$ da. $\times 24 = \frac{1}{66}$ hr.: $\frac{1}{66}$ hr. $\times 60 = \frac{60}{66} = \frac{10}{11}$ min., *Ans.*

(9.) $\frac{3}{320}$ bu. $\times 4 = \frac{3}{80}$ pk.: $\frac{3}{80}$ pk. $\times 8 = \frac{3}{10}$ qt.: $\frac{3}{10}$ qt. $\times 2 = \frac{3}{5}$ pt., *Ans.*

Art. 125.

(2.) $\frac{4}{5}$ mi. $\times 320 = \frac{1280}{5}$ rd. $= 256$ rd., *Ans.*

3.) $\$\frac{3}{5} \times 100 = \frac{300}{5}$ ct. $= 60$ ct., *Ans.*

(4.) $\frac{2}{5}$ mi. $\times 320 = \frac{640}{5}$ rd. $= 128$ rd., *Ans.*

(5.) $\frac{4}{5}$ lb. $\times 12 = \frac{48}{5}$ oz. $= 9\frac{3}{5}$ oz.: $\frac{3}{5}$ oz. $\times 20 = \frac{60}{5}$ pwt. $= 12$ pwt. *Ans.* 9 oz. 12 pwt.

(6.) $\frac{7}{16}$ T. $\times 20 = \frac{140}{16} = 8\frac{3}{4}$ cwt.; $\frac{3}{4}$ cwt. $\times 100 = \frac{300}{4}$ lb. $= 75$ lb. *Ans.* 8 cwt. 75 lb.

(7.) $\frac{5}{8}$ A. $\times\, 160 = \frac{800}{8}$ sq. rd. $= 100$ sq. rd., *Ans.*

(8.) $\frac{1}{8}$ of 63 gal. $= 7\frac{7}{8}$, and $\frac{7}{8} = 55\frac{1}{8}$ gal.: $\frac{1}{8}$ gal. $\times\, 4 = \frac{4}{8}$ or $\frac{1}{2}$ qt.: $\frac{1}{2}$ qt. $\times\, 2 = \frac{2}{2}$ or 1 pt. *Ans.* 55 gal. 1 pt.

Art. 126.

(2.) $\frac{\cancel{4}}{5} \times \frac{1}{8} \times \frac{1}{\cancel{4}} = \frac{1}{40}$ bu., *Ans.*

(3.) $\frac{4}{5} \times \frac{2}{33} = \frac{8}{165}$ rd., *Ans.* ($16\frac{1}{2}$ ft. in a rd. $= \frac{33}{2}$ ft.)

(4.) $\frac{3}{80} \times \frac{1}{16} = \frac{3}{1280}$ lb., *Ans.*

(5.) $\frac{\cancel{4}}{9} \times \frac{1}{100} \times \frac{1}{\underset{5}{\cancel{20}}} = \frac{1}{4500}$ T., *Ans.*

(6.) $\frac{3}{5} \times \frac{1}{2} \times \frac{1}{8} \times \frac{1}{4} = \frac{3}{320}$ bu., *Ans.*

(7.) $\frac{\cancel{4}}{7} \times \frac{1}{\underset{4}{\cancel{16}}} \times \frac{1}{100} = \frac{1}{2800}$ cwt., *Ans.*

(8.) $\frac{\cancel{3}}{\underset{2}{\cancel{4}}} \times \frac{1}{12} \times \frac{\cancel{2}}{\underset{11}{\cancel{33}}} = \frac{1}{264}$ rd., *Ans.*

(9.) $\frac{\cancel{8}}{9} \times \frac{1}{60} \times \frac{1}{\underset{3}{\cancel{24}}} = \frac{1}{1620}$ da., *Ans.*

(10.) $\frac{\cancel{5}}{112} \times \frac{1}{16} \times \frac{1}{\underset{20}{\cancel{100}}} = \frac{1}{35840}$ cwt., *Ans.*

Art. 127.

(2.) 2 ft. 6 in. $= 30$ in.: 6 ft. 8 in. $= 80$ in.: $\frac{30}{80} = \frac{3}{8}$, *Ans.*

(3.) 2 pk. 4 qt. $= 20$ qt.: 1 bu. $= 32$ qt.: $\frac{20}{32} = \frac{5}{8}$, *Ans.*

(4.) 2 yd. 9 in. $=81$ in.: 8 yd. 2 ft. 3 in. $=315$ in.: $\frac{81}{315}=\frac{9}{35}$, *Ans.*

(5.) 13 hr. 30 min. $=810$ min.: 1 da. $\times 24 \times 60 = 1440$ min.: $\frac{810}{1440}=\frac{9}{16}$, *Ans.*

(6.) $\frac{145}{320}=\frac{29}{64}$, *Ans.*

(7.) 2 ft. 8 in. $=32$ in.: 1 yd. $=36$ in.: $\frac{32}{36}=\frac{8}{9}$, *Ans.*

(8.) 15 mi. 123 rd. $=4923$ rd.: 35 mi. 287 rd. $=11487$ rd.: $\frac{4923}{11487}=\frac{3}{7}$, *Ans.*

(9.) 37 A. 94 sq. rd. $=6014$ sq. rd.: 168 A. 28 sq. rd. $=26908$ sq. rd.: $\frac{6014}{26908}=\frac{97}{434}$, *Ans.*

(10.) $\frac{\overset{4}{\cancel{64}}}{9}$ oz. $\times \frac{1}{\cancel{16}}=\frac{4}{9}$, *Ans.*

(11.) 2 qt. $1\frac{1}{3}$ pt. $=5\frac{1}{3}$ or $\frac{16}{3}$ pt.: 1 bu. 1 qt. $1\frac{2}{3}$ pt. $=67\frac{2}{3}$ or $\frac{203}{3}$ pt.: $\frac{16}{\cancel{3}} \times \frac{\cancel{3}}{203}=\frac{16}{203}$, *Ans.*

(12.) 1 yd. 1 ft. $1\frac{9}{11}$ in. $=49\frac{9}{11}$ in. $=\frac{548}{11}$: 3 yd. 2 ft. $8\frac{6}{7}$ in. $=140\frac{6}{7}=\frac{986}{7}$: $\frac{548}{11}\times\frac{7}{986}=\frac{3836}{10890}=\frac{1918}{5423}$, *Ans.*

Art. 128.

(3)

	hr.	min.
$\frac{2}{3}$ da. =	16	0
$\frac{3}{4}$ hr. =		45
Ans.	16	45

(4)

	da.	hr.	min.
$\frac{1}{4}$ wk. =	1	18	0
$\frac{1}{4}$ da. =		6	0
$\frac{1}{4}$ hr. =			15
Ans.	2	0	15

(5)

	da.	hr.	min.	sec.
$\frac{2}{3}$ wk. =	4	16	0	0
$\frac{5}{9}$ da. =		13	20	0
$\frac{2}{3}$ hr. =			40	0
$\frac{2}{3}$ min. =				40
Ans.	5	6	0	40

(6)

	qt.	pt.	gi.
$\frac{11}{12}$ gal. =	3	1	$1\frac{1}{3}$
$\frac{1}{12}$ qt. =		0	$\frac{2}{3}$
Ans.	3	1	2

(7)

	hr.	min.	sec.
$\frac{7}{9}$ da. =	18	40	0
$\frac{1}{18}$ hr. =		3	20
Ans.	18	36	40

(8)

	ct.
$\$\frac{5}{8}$ =	$62\frac{1}{2}$
$\$\frac{3}{40}$ =	$7\frac{1}{2}$
Ans.	55

(9.) $\frac{3}{8}$ lb. $= 6$ oz.: 6 oz. $- \frac{7}{8}$ oz. $= 5\frac{1}{8}$ oz., *Ans.*

(10.) $\frac{1}{7}$ da. $= \frac{24}{7}$ hr.: $\frac{24}{7} - \frac{6}{7} = \frac{18}{7} = 2\frac{4}{7}$ hr.: $\frac{4}{7}$ hr. $\times 60 = \frac{240}{7}$ or $34\frac{2}{7}$ min.: $\frac{2}{7}$ min. $\times 60 = \frac{120}{7}$ or $17\frac{1}{7}$ sec. *Ans.* 2 hr. 34 min. $17\frac{1}{7}$ sec.

Promiscuous Examples.

Art. 129.

(1.) $\frac{32989}{56981} = \frac{2999 \times 11}{2999 \times 19} = \frac{11}{19}$, *Ans.*

(2.) $2 + 3 = 5$: $\frac{1}{2} + \frac{2}{3} + \frac{5}{14} + \frac{8}{21} = \frac{21}{42} + \frac{28}{42} + \frac{15}{42} + \frac{16}{42} = \frac{80}{42} = \frac{40}{21} = 1\frac{19}{21}$: $5 + 1\frac{19}{21} = 6\frac{19}{21}$, *Ans.*

(3.) $\frac{25}{7} = \frac{125}{35}$: $\frac{9}{5} = \frac{63}{35}$: $\frac{125-63}{35} = \frac{62}{35} = 1\frac{27}{35}$, *Ans.*

(4.) $3\frac{5}{8} = \frac{29}{8}$: $\frac{1}{3}$ of $3\frac{1}{2} = \frac{1}{3}$ of $\frac{7}{2} = \frac{7}{6}$: $\frac{29}{8} - \frac{7}{6} = \frac{87}{24}$ $\frac{28}{24} = \frac{59}{24} = 2\frac{11}{24}$, *Ans.*

(5.) $\frac{\not{5}}{9}$ of $\frac{7}{\not{10}_{2}} = \frac{7}{18}$: $\frac{\not{2}}{5}$ of $\frac{7}{\not{12}_{6}} = \frac{7}{30}$: $\frac{7}{18} + \frac{7}{30} = \frac{35}{90} + \frac{21}{90} = \frac{56}{90} = \frac{28}{45}$, *Ans.*

(6.)

$$1\tfrac{3}{4} \div 2\tfrac{1}{2} = \frac{7}{\underset{2}{\cancel{4}}} \times \frac{\cancel{2}}{5} = \tfrac{7}{10}: \quad 5\tfrac{1}{2} \div 3\tfrac{1}{8} = \frac{11}{\cancel{2}} \times \frac{\overset{4}{\cancel{8}}}{25} = \tfrac{44}{25}:$$

$\tfrac{7}{10} + \tfrac{44}{25} = \tfrac{35}{50} + \tfrac{88}{50} = \tfrac{123}{50} = 2\tfrac{23}{50}$, *Ans.*

(7.) $10 \times \tfrac{3}{5} = \tfrac{30}{5} = 6$, *Ans.*

(8.) $10 \div \tfrac{3}{5} = 10 \times \tfrac{5}{3} = \tfrac{50}{3} = 16\tfrac{2}{3}$, *Ans.*

(9.) Any number less $\tfrac{3}{7} = \tfrac{4}{7}$: then 16 is $\tfrac{4}{7}$ of the number: 4 is $\tfrac{1}{7}$, and 28 is $\tfrac{7}{7}$, the number.

(10.) Any number plus $\tfrac{3}{7} = \tfrac{10}{7}$: then $20 = \tfrac{10}{7}$: $\tfrac{1}{7} = \tfrac{1}{10}$ of $20 = 2$: $\tfrac{7}{7} = 14$, the number.

(11.) $\tfrac{1}{3}$ of $\tfrac{5}{8} = \tfrac{5}{24}$, and $\tfrac{5}{8} - \tfrac{5}{24} = \tfrac{15}{24} - \tfrac{5}{24} = \tfrac{10}{24} = \tfrac{5}{12}$, part left.

Or, the part left may be found thus: If he sell $\tfrac{1}{3}$ of his share, he has $\tfrac{2}{3}$ of it left, and $\tfrac{2}{3}$ of $\tfrac{5}{8} = \tfrac{10}{24} = \tfrac{5}{12}$. $\tfrac{5}{12}$ of $\$900 = \tfrac{4500}{12} = \$375$, *Ans.*

(12.) I sell $\tfrac{1}{3}$ of $\tfrac{7}{12}$ of the ship $= \tfrac{7}{36}$ of the ship for $\$1944\tfrac{4}{9}$; at that rate, $\tfrac{1}{36}$ of the ship is worth $\tfrac{1}{7}$ of $\$1944\tfrac{4}{9} = \$277\tfrac{4}{9}$, and $\tfrac{36}{36}$ is worth 36 times $\$277\tfrac{4}{9} = \$10000$.

(13.) $\tfrac{2}{3}$ of $2 = \tfrac{4}{3} = 1\tfrac{1}{3}$: $\dfrac{1\frac{1}{3}}{3} = \tfrac{4}{3} \times \tfrac{1}{3} = \tfrac{4}{9}$, *Ans.*

(14.) $\tfrac{176}{368} = \dfrac{16 \times 11}{16 \times 23} = \tfrac{11}{23}$, *Ans.*

(15.) $\tfrac{1}{8} + \tfrac{1}{18} + \tfrac{13}{111} = \tfrac{333}{2664} + \tfrac{148}{2664} + \tfrac{312}{2664} = \tfrac{793}{2664}$: $\tfrac{25}{37} - \tfrac{793}{2664} = \tfrac{1800}{2664} - \tfrac{793}{2664} = \tfrac{1007}{2664}$, *Ans.*

(16.)

$$4\tfrac{9}{14} = \tfrac{65}{14}: \quad \frac{\cancel{3}}{\underset{2}{\cancel{10}}} \text{ of } \frac{\cancel{7}}{\underset{4}{\cancel{12}}} \text{ of } \frac{\overset{13}{\cancel{65}}}{\underset{2}{\cancel{14}}} = \tfrac{13}{16}: \quad 1 - \tfrac{13}{16} = \tfrac{3}{16}, \textit{ Ans.}$$

(17.) $\frac{2}{3} \div \frac{5}{7} = \frac{2}{3} \times \frac{7}{5} = \frac{14}{15}$: $\frac{5}{8} \div \frac{10}{11} = \frac{\cancel{5}}{8} \times \frac{11}{\underset{2}{\cancel{10}}} = \frac{11}{16}$: $\frac{14}{15} - \frac{11}{16} = \frac{224}{240} - \frac{165}{240} = \frac{59}{240}$, *Ans.*

(18.) In $\frac{1}{15}$ of an hour he walks $\frac{1}{7}$ of 2044 rd., which is 292 rd.: $1\frac{14}{15} = \frac{29}{15}$: in $\frac{29}{15}$ hr. he will walk 29 times 292 rd. = 8468 rd., *Ans.*

(19.) $1\frac{1}{4}$ ft. = 15 in. = $\frac{45}{3}$: $3\frac{1}{3} = \frac{10}{3}$: $\frac{10}{45} = \frac{2}{9}$, *Ans.*

(20.) $3\frac{1}{5} + 3\frac{2}{3} = \frac{16}{5} + \frac{11}{3} = \frac{48}{15} + \frac{55}{15} = \frac{103}{15}$. *Ans.* $\frac{48}{103}$ and $\frac{55}{103}$.

(21.) $\frac{5}{8}$ of \$2400 = \$1500: \$1500 + \$500 = \$2000. If $\frac{5}{4}$ of B's money = \$2000, $\frac{1}{4}$ is $\frac{1}{5}$ of \$2000, which is \$400, and the whole will be 4 times \$400, which are \$1600, *Ans.*

(22.) If \$2200 are $\frac{5}{12}$ of the elder one's share, $\frac{1}{12}$ is \$440, and $\frac{12}{12}$, the elder one's share = \$5280; if \$5280 are $\frac{16}{35}$ of the whole estate, $\frac{1}{35}$ is \$330, and $\frac{35}{35}$ = \$11550; \$2200 + \$5280 = \$7480; \$11550 — \$7480 = \$4070; each daughter had $\frac{1}{3}$ of \$4070 = \$1356$\frac{2}{3}$, *Ans.*

PRACTICE.

Art. 130.

(3.) $12\frac{1}{2} = \frac{25}{2}$: $18\frac{3}{4}$ ct. = \$$\frac{3}{16}$: $\frac{25}{2} \times \frac{3}{16}$ = \$$\frac{75}{32}$ = \$2.34$\frac{3}{8}$, *Ans.*

(4.)

$$\$2.25 = \$2\tfrac{1}{4} = \frac{\overset{3}{\cancel{9}}}{\cancel{4}} \times \frac{\overset{4}{\cancel{16}}}{\cancel{3}} = 12 \text{ yd., } \textit{Ans.}$$

(5.) $\frac{11}{2} \times \frac{5}{8} = \frac{55}{16}$: \$$\frac{1}{16} = 6\frac{1}{4}$ ct.; \$$\frac{55}{16}$ = \$3.43$\frac{3}{4}$, *Ans.*

(6.) $\$66.25 = \frac{265}{4}$: $\$3.75 = 3\frac{3}{4} = \frac{15}{4}$: $\frac{\overset{53}{\cancel{265}}}{\cancel{4}} \times \frac{\cancel{4}}{\underset{3}{\cancel{15}}} = \frac{53}{3}$ $= 17\frac{2}{3}$ doz., *Ans.*

(7.) $\$2.37\frac{1}{2} = 2\frac{3}{8} = \frac{19}{8}$: $\frac{19}{\cancel{8}} \times \frac{\overset{10}{\cancel{80}}}{1} = \$190$, *Ans.*

(8.) $\$4.87\frac{1}{2} = 4\frac{7}{8} = \frac{39}{8}$: $\frac{\cancel{39}}{1} \times \frac{8}{\cancel{39}} = 8$ men, *Ans.*

(9.) $\$8.33\frac{1}{3} = 8\frac{1}{3} = \frac{25}{3}$: $\frac{25}{\cancel{3}} \times \frac{\overset{12}{\cancel{36}}}{1} = \$300$, *Ans.*

(10.) $\$246.66\frac{2}{3} = 246\frac{2}{3} = \frac{740}{3}$: $\$1.33\frac{1}{3} = 1\frac{1}{3} = \frac{4}{3}$: $\frac{\overset{185}{\cancel{740}}}{\cancel{3}} \times \frac{\cancel{3}}{\cancel{4}} = 185$ yd., *Ans.*

(12.) $\$18\frac{1}{3} = \frac{55}{3}$: $\$229\frac{1}{6} = \frac{1375}{6}$: $\frac{\overset{275}{\cancel{1375}}}{\underset{2}{\cancel{6}}} \times \frac{3}{\underset{11}{\cancel{55}}} = \frac{275}{22} =$ $12\frac{1}{2}$, *Ans.*

(13.) 120 sq. rd. $= \frac{3}{4}$ A. $\$125.60 \times 11 = \$1381.60 =$ cost of 11 A.: $\frac{3}{4}$ of $\$125.60 = \$94.20 =$ cost of 120 sq. rd.: $\$1381.60 + \$94.20 = \$1475.80$, *Ans.*

(14.) $\frac{100\cancel{0}}{25\cancel{0}} = 40$ lots. 50 ft. $\times$ 150 ft. $= 7500$ sq. ft.: 7500 sq. ft. $\times 40 = 300000$ sq. ft., $\div 9 = 33333$ sq. yd. $+ 3$ sq. ft.: 33333 sq. yd. $\div 30\frac{1}{4} = 1101$ sq. rd. $+ 27\frac{3}{4}$ sq. yd.: $\frac{3}{4}$ sq. yd. $= \frac{27}{4}$ sq. ft. $= 6$ sq. ft. $+ \frac{3}{4}$; $\frac{3}{4}$ sq. ft. $\times 144 = 108$ sq. in.; 6 sq. ft. $+ 3$ sq. ft. $= 9$ sq. ft. $= 1$ sq. yd., which added to $27 = 28$ sq. yd.: 1101 sq. rd. $\div 160 = 6$ A. 141 sq. rd. *Ans.* 6 A. 141 sq. rd. 28 sq. yd. 108 sq. in.

(15.) 2 qt. $= \frac{1}{4}$ pk. $= \frac{1}{16}$ bu.: 3 pk. $= \frac{3}{4}$ or $\frac{12}{16}$ bu.: \$6.20 $= \$6\frac{1}{5} = \frac{31}{5}$: $83\frac{13}{16} = \frac{1341}{16}$, $\times \frac{31}{5} = \frac{41571}{80} = \$519.63\frac{3}{4}$, *Ans.*

(16.)

$$167\frac{1}{2} = \frac{335}{2}: \quad \frac{335}{\cancel{2}} \times \frac{\overset{2}{\cancel{4}}}{3} = \frac{670}{3} = 223\frac{1}{3} \text{ bu.}: \tfrac{1}{3} \text{ bu.} = \tfrac{4}{3}$$

or $1\frac{1}{3}$ pk.: $\frac{1}{3}$ pk. $= \frac{8}{3}$ or $2\frac{2}{3}$ qt.: $\frac{2}{3}$ qt. $= \frac{4}{3}$ or $1\frac{1}{3}$ pt. *Ans.* 223 bu. 1 pk. 2 qt. $1\frac{1}{3}$ pt.

(17.) \$1.75 $= 1\frac{3}{4}$ or $\frac{7}{4}$: $\frac{7}{2} \times \frac{7}{4} = \frac{49}{8} = \$6.12\frac{1}{2}$, *Ans.*

(18.)

$$\$1.50 = \tfrac{3}{2};\ \$7.12\tfrac{1}{2} = \tfrac{57}{8}: \quad \frac{\overset{19}{\cancel{57}}}{\underset{4}{\cancel{8}}} \times \frac{\cancel{2}}{\cancel{3}} = \frac{19}{4} = 4\tfrac{3}{4} \text{ yd., } \textit{Ans.}$$

(19.) 12 oz. $= \frac{12}{16}$ or $\frac{3}{4}$ lb.: $45\frac{3}{4}$ lb. $= \frac{183}{4}$: $\frac{183}{4} \times \frac{3}{8} = \frac{549}{32} = \$17.15\frac{5}{8}$, *Ans.*

(20.) $\$0.93\frac{3}{4} = \left\{ \begin{matrix} 87\frac{1}{2} = \frac{7}{8} = \frac{14}{16} \\ 6\frac{1}{4} \quad\quad = \frac{1}{16} \end{matrix} \right\} = \frac{15}{16}$: $\$2\frac{15}{16} = \$\frac{47}{16}$:

$$\frac{47}{\underset{2}{\cancel{16}}} \times \frac{\cancel{8}}{1} = \frac{47}{2} = 23\tfrac{1}{2} \text{ lb., } \textit{Ans.}$$

(21.) 2 T. 9 cwt. $= 49$ cwt.: $37\frac{1}{2}$ ct. per lb. $= \$37\frac{1}{2}$ per cwt.: $\frac{49}{1} \times \frac{75}{2} = \$1837.50$, *Ans.*

(22.)

$$\$3.90 = 3\tfrac{9}{10} = \tfrac{39}{10}: \quad \frac{\overset{\overset{5}{\cancel{10}}}{\cancel{100}}}{\underset{\underset{2}{\cancel{4}}}{\cancel{12}}} \times \frac{\overset{13}{\cancel{39}}}{\cancel{10}} = \$32.50, \textit{ Ans.}$$

(23.)

$$3\tfrac{3}{4} = \tfrac{15}{4};\ \$5.40 = 5\tfrac{4}{10} = \tfrac{54}{10}: \quad \frac{\overset{3}{\cancel{15}}}{\underset{2}{\cancel{4}}} \times \frac{\overset{27}{\cancel{54}}}{\underset{2}{\cancel{10}}} = \$20.25, \textit{ Ans.}$$

(24.)

$$\left.\begin{array}{r}\dfrac{13}{\not 2}\times\dfrac{1}{\not 3}\times\dfrac{\not 12^{\,\not 4^{\,2}}\text{ (1 doz.)}}{1}=\$26.\\ \frac{75}{2}\times\frac{3}{8}=\frac{225}{16}=\$14.06\frac{1}{4}\end{array}\right\}=\$40.06\tfrac{1}{4}:$$

$$\$40.06\tfrac{1}{4}-\$36=\$4.06\tfrac{1}{4}=\$4\tfrac{1}{16}=\$\tfrac{65}{16}:\ \frac{65}{\not 16_{\,2}}\times\frac{\not 8}{1}=\tfrac{65}{2}=$$

$32\frac{1}{2}$ lb., *Ans.*

DECIMAL FRACTIONS.

Art. 135.

REMARKS.—Pupils must have a thorough knowledge of common fractions, before they can understand fully the reason of the rules in decimals.

When a pupil is in doubt with regard to the accuracy of the result in any operation involving decimals, let him convert the decimals into common fractions, and then perform the work; the results in both cases ought to be the same.

It is a useful exercise to perform the same operations in equivalent common and decimal fractions. Thus, they may be required to perform the operations indicated in the following examples, by the rules for common fractions; then to convert the common fractions into decimals, and work by the rules for decimals.

(5.) .26
(6.) .35
(7.) .87
(8.) 4.19
(9.) .005
(10.) .054
(11.) .304
(12.) 7.293
(13.) 25.047
(14.) .0205
(15.) .4125

(16.) .00009
(17.) .900
(18.) .00605
(19.) .20304
(20.) .000007
(21.) .000203
(22.) .300004
(23.) .0000024
(24.) .0080006
(25.) .000200
(26.) .00000002

(27.) .00000907
(28.) .20020003
(29.) 1.010100
(30.) .01010001
(31.) 106.037
(32.) 1000.001
(33.) .225
(34.) 200.025
(35.) .002929
(36.) 2900.000029
(37.) .001000005
(38.) .0000000202
(39.) 200.0000000002
(40.) 65.006005
(41.) .3 .7 .09 .17 .23 .41 .53
(42.) .87 .97. .123. .289 .487 .733
(43.) .003 .0101 .00053 .000503

Art. 136.

(4.) Twenty-eight *thousandths;* three hundred and forty-one *thousandths;* two and three hundred and twenty-seven *thousandths;* fifty and five *thousandths;* one hundred and eighty-four and one hundred and seventy-three *thousandths.*

(5.) Three *ten-thousandths;* six hundred and twenty-five *ten-thousandths;* two thousand three hundred and seventy-four *ten-thousandths;* two thousand and six *ten-thousandths;* one hundred and four *ten-thousandths.*

(6.) Three and two hundred and five *ten-thousandths;* eight hundred and ten and two thousand four hundred and six *ten-thousandths;* ten thousand seven hundred and twenty and nine hundred and five *ten-thousandths.*

(7.) Four *hundred-thousandths;* one hundred and thirty-seven *hundred-thousandths;* two thousand three hundred and seventy-six *hundred-thousandths;* one thousand and seven *hundred-thousandths.*

(8.) One thousand seven hundred and sixty-eight *millionths;* forty thousand and thirty-five *millionths;* seventy and three hundred and sixty thousand and four *millionths.*

(9.) One million ten thousand one hundred and one *ten-millionths;* forty thousand and five *hundred-millionths;* one hundred thousand three hundred and four *hundred-millionths.*

(10.) Thirty-one thousand four hundred and fifty-six *hundred-thousandths;* one hundred and thirty-three *millionths;* sixty and four *hundredths;* forty-five and one thousand and three *ten-thousandths.*

(11.) Three hundred and fifty-seven and seventy-five *hundredths;* four thousand nine hundred and twenty-eight *ten-thousandths;* five and nine hundred and forty-five *thousandths;* six hundred and eighty-one and two *ten-thousandths.*

(12.) Seventy and one million two hundred thousand seven hundred and sixty-four *ten-millionths;* nine hundred and fifty-four and two hundred and three *thousandths;* thirty-eight and twenty-seven *thousandths.*

(13.) One thousand and seven and three thousand one hundred and fifty-four *ten-thousandths;* seven thousand four hundred and ninety-six and thirty-five million four hundred and ninety-one thousand seven hundred and sixty-eight *hundred-millionths.*

(14.) Seven hundred and fifteen *hundred-thousandths;* three and five *hundred-thousandths;* twenty-eight and ten million sixty-five thousand seven hundred and one *hundred-millionths.*

(15.) Thirteen and eight trillion two hundred and forty-one billion ninety-four million seven hundred and ten thousand nine hundred and forty-seven *ten-quadrillionths.*

(16.) $\frac{9}{10}$; $\frac{13}{100}$; $\frac{19}{100}$; etc.

(17.) $\frac{91}{100}$; $\frac{347}{1000}$; $\frac{513}{1000}$; etc.

(18.) $\frac{7}{1000}$; $\frac{207}{10000}$; $\frac{79}{100000}$; $\frac{1007}{1000000}$.

(19.) $1\frac{36}{100}$; $\frac{3421}{10000}$; $\frac{3401}{100000}$; $\frac{900}{10000}$.

(20.) $\frac{1}{1000}$; $\frac{5302}{10000}$; $8\frac{1}{100}$; $\frac{53}{1000000}$.

Art. 141.

(2.) $.6 = \frac{6}{10} = \frac{3}{5}$, *Ans.* (3.) $.25 = \frac{25}{100} = \frac{1}{4}$, *Ans.*

(4.) $.375 = \frac{375}{1000} = \frac{3}{8}$, *Ans.*

(5.) $.035 = \frac{35}{1000} = \frac{7}{200}$, *Ans.*

(6.) $.5625 = \frac{5625}{10000} = \frac{9}{16}$, *Ans.*

(7.) $.34375 = \frac{34375}{100000} = \frac{11}{32}$, *Ans.*

(8.) $.1484375 = \frac{1484375}{10000000} = \frac{19}{128}$, *Ans.*

(9.) $4.02 = 4\frac{2}{100} = 4\frac{1}{50}$, *Ans.*

(10.) $8.415 = 8\frac{415}{1000} = 8\frac{83}{200}$, *Ans.*

Art. 142.

(2.) $\frac{4}{5} = \frac{4.0}{5} = .8$, *Ans.* (3.) $\frac{5}{8} = \frac{5.000}{8} = .625$, *Ans.*

(4.) $\frac{7}{25} = \frac{7.00}{25} = \frac{1.40}{5} = .28$, *Ans.*

(5.) $\frac{3}{40} = \frac{3.000}{40} = .075$, *Ans.*

(6.) $\frac{15}{16} = \frac{15.0000}{16} = .9375$, *Ans.*

(7.) $\frac{1}{1250} = \frac{1.0000}{1250} = .0008$, *Ans.*

(8.) $\frac{9}{400} = \frac{9.0000}{400} = .0225$, *Ans.*

(9.) $\frac{1}{256} = \frac{1.00000000}{256} = .00390625$, *Ans.*

(10.) $\frac{5}{6} = \frac{5.0000}{6} = .8333+$, *Ans.*

(11.) $\frac{1}{11} = \frac{1.000000}{11} = .090909+$, *Ans.*

(12.) $\frac{4}{33} = \frac{4.000000}{33} = .121212+$, *Ans.*

Art. 143.

(2)
37.1065
432.07
4.20733
11.706
485.08983

(3)
4.0004
28.035
8.07
.09404
40.19944

(4)
3.25
6.4
.35
10.00

(5)
21.611
6888.32
3.4167
6913.3477

(6)
6.61
636.1
6516.14
67.1234
5.1233
7231.0967

(7)
4.8
43.31
74.019
11.204
133.333

(8)
45.019
7.00071
93.4327
6.0401
151.49251

(9)
432.432
61.0793
100.07794
6.009
1000.1001
1599.69834

(10)
16.041
9.000094
33.27
8.969
32.719906
100.000000

(11)
204.0009
103.00000009
42.009099
430.99
220.0000009
999.99999999

(12)
.0035
.00035
.000035
.0000035
.0038885

Art. 144.

(2)
97.5168
38.25942
59.25738

(3)
20.014
7.0021
13.0119

(4)
5.03
2.115
2.915

(5)
24.0042
13.7013
10.3029

(6)
170.0035
68.00181
102.00169

(7)
.0142
.005
.0092

(8)
.05
.0024
.0476

(9)
13.5
8.037
5.463

(10)	(11)	(12)
3.00000	29.0029	5.000
.00003	19.003	.125
2.99997	9.9999	4.875

(13)	(14)	(15)
1000.0000	1.000000	.025
.0001	.000001	.000025
999.9999	.999999	.024975

Art. 147.

(4)	(5)	(6)	(7)
33.21	32.16	.125	.35
4.41	22.5	9	7
3321	16080	1.125	2.45
13284	6432		
13284	6432		
146.4561	723.600		

(10.) $.15 \times .7 = \frac{15}{100} \times \frac{7}{10} = \frac{105}{1000} = .105$, *Ans.*

(13.) $1.035 \times 17 = 17.595$, *Ans.*

(14)	(15)	(16)	(17)
19	4.5	.625	61.76
.125	4	64	.0071
95	18.0	2500	6176
38		3750	43232
19		40.000	.438496
2.375			

(18)
```
   1.325
   .0716
  ------
    7950
   1325
  9275
 -------
.0948700
```

(24)
```
  .1
 .01
----
.001
```

(25)
```
   100
.0001
------
00.0100 = .01, Ans.
```

(26)
```
    .043
   .0021
 -------
      43
     86
 -------
.0000903
```

(27)
```
  40000
.000001
-------
.040000
```

(28)
```
   .09375
    1.064
  -------
    37500
   56250
  93750
---------
.09975000
```

Art. 150.

Suggestions to Teachers.—The division of decimals is generally a troublesome subject to pupils; this arises from a want of attention to the rule. Should the pupil be at a loss to understand why, in some cases, when the divisor and dividend are both decimals, the quotient should be a whole number, let him read the remarks on the division of fractions, page 120. When the divisor contains more decimal places than the dividend, it is best, before commencing the division, to reduce them both to the same denomination, that is, to make the number of decimal places the same in both; the quotient will then be a whole number.

(7)
```
.03)1.125
    -----
     37.5, Ans.
```

(8)
```
27.5)86.075(3.13, Ans.
     825
     ---
      357
      275
      ---
       825
       825
       ---
```

(9)
```
3.44)24.73704(7.191, Ans.
     2408
     ----
       657
       344
       ----
       3130
       3096
       ----
         344
         344
         ---
```

```
(10)
4.123)206.166492(50.004, Ans.
      20615
      -----
        16492
        16492
        -----
```

```
(13)
.5)21.0(42. Ans.
   20
   --
   10
   10
   --
```

```
(14)
.008)2.000
     -----
       250, Ans.
```

```
(15)
5)37.20
  -----
   7.44, Ans.
```

```
(16)
454)100.8788(.2222, Ans.
    908
    ---
    1007
     908
    ----
      998
      908
      ---
       908
       908
       ---
```

```
(18)
.108649)9811.004700(90300, Ans.
        977841
        ------
         325947
         325947
         ------
             00
```

```
(19)
.19).21318(1.122, Ans.
     19
     --
      23
      19
      --
       41
       38
       --
        38
        38
        --
```

```
(20)
.3189)102048.0000(320000, Ans.
      9567
      ----
       6378
       6378
       ----
       0000
```

```
(21)
3189).102048(.000032, Ans.
       9567
       ----
        6378
        6378
        ----
```

```
(22)
.0225)9.9000(440, Ans.
      900
      ---
       900
       900
       ---
         0
```

(26)

$$10\overline{).10}$$
.01, *Ans.*

(27)

$$.1\overline{)1.0}$$
10, *Ans.*

(28)

$$.01\overline{)10.00}$$
1000, *Ans.*

(29.) $\frac{1.7}{64}=\frac{1.7000000}{64}=.0265625$, *Ans.*

(30)

$$80\overline{).080}$$
.001, *Ans.*

(31)

$$7\overline{)1.5000000}$$
.2142857+, *Ans.*

(32)

```
32.76)11.100000000(.3388278+ Ans.
       9828
       12720
        9828
        28920
        26208
         27120
         26208
           9120
           6552
           25680
           22932
            27480
            26208
```

(33)

```
3.21).0123000000(.00383177+
       963                Ans.
       2670
       2568
        1020
         963
          570
          321
          2490
          2247
           2430
           2247
```

Art. 151.

(2.) .035 pk. × 8 = .280 qt.: .28 qt. × 2 = .56 pt., *Ans.*

(3.) .0075 bu. × 4 = .0300 pk.: .03 pk. × 8 = .24 qt., *Ans.*

(4.) .005 yd. × 3 = .015 ft., × 12 = 0.180 in. = .18 in., *Ans.*

(5.) .00546875 A. × 160 = 00.87500000 sq. rd. = .875 sq. rd., *Ans.*

Art. 152.

(2.) .75 yd. $\times$ 3 = 2.25 ft.: .25 ft. $\times$ 12 = 3.00 in. *Ans.* 2 ft. 3 in.

(3.) .3375 A. $\times$ 160 = 54.0000 sq. rd. = 54 sq. rd., *Ans.*

(4.) .7 lb. $\times$ 12 = 8.4 oz.: .4 oz. $\times$ 20 = 8.0 pwt. = 8 pwt. *Ans.* 8 oz. 8 pwt.

(5.) .8125 bu. $\times$ 4 = 3.2500 pk.: .25 pk. $\times$ 8 = 2.00 qt. = 2 qt. *Ans.* 3 pk. 2 qt.

(6.) .44 mi. $\times$ 320 = 140.8 rd.: .8 rd. $\times$ $5\frac{1}{2}$ = 4.4 yd.: .4 yd. $\times$ 3 = 1.2 ft.: .2 ft. $\times$ 12 = 2.4 in. *Ans.* 140 rd. 4 yd. 1 ft. 2.4 in.

(7.) .33625 cwt. $\times$ 100 = 33.625 lb.: .625 lb. $\times$ 16 = 10.000 oz. = 10 oz. *Ans.* 33 lb. 10 oz.

Art. 153.

(2.) .72 qt. $\div$ 8 = .09 pk., $\div$ 4 = .0225 bu., *Ans.*

(3.) .77 yd. $\div$ $5\frac{1}{2}$ = .14 rd., $\div$ 320 = .0004375 mi., *Ans.*

(4.) .25 pt. $\div$ 2 = .125 qt., $\div$ 4 = .03125 gal., *Ans.*

(5.) .6 pt. $\div$ 2 = .3 qt., $\div$ 8 = .0375 pk. $\div$ 4 = .009375 bu., *Ans.*

(6.) .7 rd. $\div$ 320 = .0021875 mi., *Ans.*

Art. 154.

(1.) \$0.40 $\times$ 9 = \$3.60: \$0.75 $\times$ 12 = \$9.00: \$3.60 + \$9.00 = \$12.60, *Ans.*

(2.) \$0.45 $\times$ 2.3 = \$1.035: \$0.375 $\times$ 1.5 = \$0.5625: \$1.035 + \$0.5625 = \$1.5975, *Ans.*

(3.) \$2.6875 $\times$ $16\frac{1}{4}$ = \$43.671875, *Ans.*

(4.) 35.25 $\div$ .75 = 47 bu., *Ans.*

(5.) $98.4 \div 2.5625 = 38.4$ yd., *Ans.*

(6.) 6 cwt. 50 lb. = 6.5 cwt.: $\$3.25 \times 6.5 = \$21.125$, *Ans.*

(7.) 14 bu. 3 pk. 4 qt. = 14.875 bu.: $\$0.625 \times 14.857 = \$9.296875$, *Ans.*

(8.) 13 A. 115 sq. rd. = 13.71875 A.: $\$17.28 \times 13.71875 = \$237.06$, *Ans.*

(9.) $\$9.296875 \div \$0.3125 = 29.75$ bu. = 29 bu. 3 pk., *Ans.*

(10.) $59.265 \div 4.32 = 13.71875$ A. = 13 A. 115 sq. rd., *Ans.*

(11.) 1 gal. would cost $\$\frac{49}{63} = \$\frac{7}{9} = \$0.77\frac{7}{9}$: $464 \times .77\frac{7}{9} = \$360.88$, *Ans.*

(12.) .34 yd. $\times$ 3 = 1.02 ft.:
.02 ft. $\times$ 12 = .24 in.:
1.07 ft.: .07 ft. $\times$ 12 = .84 in.

	ft.	.in
	1	.24
	1	.84
		8.92
Ans.	2	10.00

(13.) .625 gal. $\times$ 4 = 2.500 qt.:
.5 qt. $\times$ 2 = 1.0 pt.:
.75 qt. $\times$ 2 = 1.5 pt.

	qt.	pt.
	2	1
		1.5
Ans.	3	.5

(14.) 1.53 yd. $\times$ 3 = 4.59 ft.:
.59 ft. $\times$ 12 = 7.08 in.

	ft.	in.
	4	7.08
	2	3.08
Ans.	2	4

(15.) $365.25 \times .05 = 18.2625$ da.: .2625 da. $\times$ 24 = 6.3 hr.: 6.3 hr. — .5 hr. = 5.8 hr.: .8 hr. $\times$ 60 = 48 min. *Ans.* 18 da. 5 hr. 48 min.

(16.) .41 da. = 9.84 hr.: 9.84 hr. — .16 hr. = 9.68 hr.: .68 hr. $\times$ 60 = 40.8 min.: .8 min. $\times$ 60 = 48 sec. *Ans.* 9 hr. 40 min. 48 sec.

(17.) 365.25 da. $\times$ .3 = 109.575 da.: .575 da. $\times$ 24 = 13.8 hr.: .8 hr. $\times$ 60 = 48 min. *Ans.* 109 da. 13 hr. 48 min.

(18.) 3 in. = $\frac{1}{4}$ ft.: $2\frac{1}{4}$ or $\frac{9}{4}$ ft. = $\frac{3}{4}$ yd.: $343\frac{3}{4}$ $\times$ \$0.16 = \$55.00, *Ans.*

(19.) 17 mi. 135 rd. = 17.421875 mi.: \$690.35 $\times$ 17.421875 = \$12027.19140625, *Ans.*

THE METRIC SYSTEM.

Art. 160.

(3.) 20 Km. $\times$ .62137 = 12.42740 mi., *Ans.*

(4.) 160 acres $\div$ 2.471 = 64.75+ Ha., *Ans.*

(5.) 49 m. $\times$ 39.37 = 1929.13 in., $\div$ 12 = 160 ft. 9.13 in.: 160 ft. $\div$ 3 = 53 yd. 1 ft.: 53 yd. $\div$ $5\frac{1}{2}$ = 9 rd. $3\frac{1}{2}$ yd.: $\frac{1}{2}$ yd. = $\frac{3}{2}$ or $1\frac{1}{2}$ ft.; $\frac{1}{2}$ ft. = 6 in.; 9.13 in. + 6 in. = 15.13 in. = 1 ft. 3.13 in.; 1 + 1 + 1 = 3 ft. = 1 yd.; 3 yd. + 1 yd. = 4 yd. *Ans.* 9 rd. 4 yd. 3.13 in.

(6.) 15 g. $\times$ 15.432 = 231.480 gr. T., $\div$ 24 = 9 pwt. 15.48 gr., *Ans.*

(7.) 42 bu. $\div$ 2.8375 = 14.8+ Hl., *Ans.*

(8.) 500 sters $\times$ .2759 = 137.95 C., *Ans.*

(9.) 9 m. $\times$ 5 m. = 4.5 m^2., $\times$ 1.196 = 5.382 sq. yd., *Ans.*

(10.) 32 l. $\times$ 1.0567 = 33.8144 qt., $\div$ 4 = 8.4536 gal., *Ans.*

Art. 161.

(1.) 127 dl. + 234.5 dl. = 361.5 dl., $\div$ 10 = 36.15 l.: 1563 cl. $\div$ 100 = 15.63 l.: 4.87 l. + 36 15 l. + 15.63 l. = 56.65 l., *Ans.*

(2.) 45 Ha. = 4500 a., @ \$3.32 = \$14940, *Ans.*

(3.) $457.92 \div 3 = 152.64$ m., *Ans.*

(4.) $.72 \times .48 \times .5 = .1728$: $\$.8640 \div .1728 = \$5$, *Ans.*

(5)	(6)	(7)
380)454.10(1.195	4685	346.75)194.1800(0.56
380 *Ans.* $1.195	1.6	173375
741	28110	208050
380	4685	208050
3610	7496.0	
3420		*Ans.* $0.56
1900	*Ans.* 7496 Hl.	
1900		

(8.) 1 M. $\times 100 = 100$ cm.: $100 \div 2 = 50$, the number of coins: 50×5 g. $= 250$ g., *Ans.*

(9.) $1.25 \times 6.5 = 8.125, \div 1.85 = 4.39+$ m., *Ans.*

(10.) 60 mi. $\div .62137 = 96.56+$ Km., *Ans.*

(11.) 29 Mm. $\times$ 22.4 Mm. $= 649.6$ Mm^2, *Ans.*

(12.) 13.24 Km. $\times 1000 = 13240$ m., $\div .715$ m. $= 18517+$, which would, of course, necessitate his taking 18518 steps, *Ans.*

NOTE.—The answer to the example given here is also 18517+ steps.

PERCENTAGE.

Art. 164.

(6)
165
$.03\frac{1}{3}$
495
55
5.50, *Ans.*

(7.) $240 \times .03\frac{3}{4} = 9$, *Ans.*

(14.) $8\frac{1}{3}\% = \frac{1}{12}$: $\frac{1}{12}$ of $384 = 32$, *Ans.*

(16.) $12\frac{1}{2}\% = \frac{1}{8}$: $\frac{1}{8}$ of $292 = 36.5$, *Ans.*

(19.)

$$18\tfrac{3}{4} = \tfrac{3}{16}:\quad \frac{\overset{.7}{\cancel{11.2}}}{1} \times \frac{3}{\cancel{16}} = 2.1, \textit{ Ans.}$$

(20.)

$$20\% = \tfrac{1}{5}:\quad \frac{\overset{1.97}{\cancel{9.85}}}{1} \times \frac{1}{\cancel{5}} = 1.97, \textit{ Ans.}$$

(21.) $25\% = \frac{1}{4}$: $\frac{1}{4}$ of $43 = 10.75$, *Ans.*

(22.) $33\frac{1}{3}\% = \frac{1}{3}$: $\frac{1}{3}$ of $6.93 = 2.31$, *Ans.*

(23.) $45 \times 5.7 = 2.565$, *Ans.*

(24.) $50\% = \frac{1}{2}$: $\frac{1}{2}$ of $38.75 = 19.375$, *Ans.*

(25.) $\frac{1}{2}\% = \frac{1}{200}$: $\frac{1}{200}$ of $456 = 2.28$, *Ans.*

(26.) $\frac{3}{8}\% = .00375$: $464 \times .00375 = 1.74$, *Ans.*

(27.) $\frac{7}{16} = .4375$: $144 \times .4375 = 63$, *Ans.*

(28.) $125\% = \frac{5}{4}$: $\frac{5}{4}$ of $36 = 45$, *Ans.*

(29.) 208% of $650 = 650 \times 2.08 = 1352$, *Ans.*

(30.) $4\frac{1}{2}$ times $12 = 48 + 6 = 54$, *Ans.*

(31.) 10 times $24.75 = 247.5$, *Ans.*

Art. 165.

(3.) 3 is $\frac{1}{5}$ of 15: $\frac{1}{5} = 20\%$, *Ans.*

(4.) 6 is $\frac{3}{25}$ of 50: $\frac{3}{25} = .12 = 12\%$, *Ans.*

(5.) 4.5 is $\frac{4.5}{75}\%$ of $75 = \frac{3}{5} = .6 = 6\%$, *Ans.*

(11)

$$243)8.505(.035 = 3\tfrac{1}{2}\%, \textit{ Ans.}$$
$$\begin{array}{l} 729 \\ \hline 1215 \\ 1215 \\ \hline \end{array}$$

(12.) $.002$ of $2 = .002 \div 2 = .001 = \frac{1}{10}$ of 1%, *Ans.*

(13.) $13.245 \div 3532 = .00375 = \frac{375}{100000} = \frac{3}{8}\%$, *Ans.*

(14.) $\frac{3}{\not{2}\not{5}_{5}} \times \frac{\not{5}}{4} = \frac{3}{20} = 15\%$, *Ans.*

(15.) $\frac{\not{2}}{\not{1}\not{5}_{5}} \times \frac{\not{3}}{\not{2}} = \frac{1}{5} = 20\%$, *Ans.*

(16.) $\frac{\not{2}}{\not{7}} \times \frac{\overset{3}{\not{2}\not{1}}}{\not{1}\not{6}_{8}} = \frac{3}{8} = 37\tfrac{1}{2}\%$, *Ans.*

(17.) $\frac{\overset{3}{\not{2}\not{1}}}{4} \times \frac{3}{\not{3}\not{5}_{5}} = \frac{9}{20} = 45\%$, *Ans.*

(18.) $\frac{\not{6}\not{5}}{\not{6}_{2}} \times \frac{\overset{3}{\not{9}}}{\not{5}\not{2}\not{0}_{8}} = \frac{3}{16} = 18\tfrac{3}{4}\%$, *Ans.*

Art. 166.

(3.) $20\% = \frac{1}{5}$: $60 \times 5 = 300$, *Ans.*

(4.) $75\% = \frac{3}{4}$. If 90 is $\frac{3}{4}$, $\frac{1}{4} = 30$, and $\frac{4}{4} = 120$, *Ans.*

(5.) $125\% = \frac{5}{4}$. If 85 is $\frac{5}{4}$, $\frac{1}{4} = 17$, and $\frac{4}{4} = 68$, *Ans.*

(6.) $7.13 \div .23 = 31$, *Ans.*

(7.) $20.23 \div .34 = 59.5$, *Ans.*

(8.) $23.5 \div .47 = 50$, *Ans.*

(9.) If 45 is $1\frac{1}{2}\%$, $\frac{1}{2}\%$ is $\frac{1}{3}$ of $45 = 15$: $1\% = 2$ times $15 = 30$: 100 times $30 = 3000$, the number.

(10.) $12\frac{1}{2}\% = \frac{1}{8}$: $2.25 \times 8 = 18$, *Ans.*

(11.) 1% is $\frac{1}{250}$ of $\frac{3}{4} = \frac{3}{1000}$: 100% is 100 times $\frac{3}{1000} = \frac{300}{1000} = \frac{3}{10}$, *Ans.*

(12.) $14\frac{2}{7} = \frac{100}{7}$: $16\frac{2}{3}\% = \frac{1}{6}$. If $\frac{100}{7} = \frac{1}{6}$, $\frac{6}{6} = \frac{600}{7} = 85\frac{5}{7}$, *Ans.*

Art. 167.

(3.) $721 \div 1.03 = 700$, *Ans.*

(4.) $100\% - 66\% = 34\%$: $68 \div .34 = 200$, *Ans.*

(5.) If $2125 = \frac{5}{4}$, $\frac{1}{4} = 425$, and $\frac{4}{4} = 1700$, *Ans.*

(6.) If $7.52 = \frac{94}{100}$, $\frac{1}{100} = \frac{7.52}{94} = 8$, and the number $= 8$, *Ans.*

(7.) $37\frac{1}{2}\% = \frac{3}{8}$. If $8250 = \frac{11}{8}$, $\frac{1}{8} = 750$, and $\frac{8}{8} = 6000$, *Ans.*

(8.) $10\% = \frac{1}{10}$, then $\frac{9}{10}$ of the fraction $= \frac{3}{8}$: $\frac{3}{8} \times \frac{10}{9} = \frac{30}{72} = \frac{5}{12}$, *Ans.*

(9.) $20\% = \frac{1}{5}$. If $6.6 = \frac{6}{5}$, $\frac{1}{5} = 1.1$, and $\frac{5}{5} = 5.5$, *Ans.*

Art. 169.

(1.) $800 \times .36 = 288.00$: $\$800 - \$288 = \$512$, *Ans.*

(2.) $300 - 225 = 75 = \frac{1}{4}$ of $300 = 25\%$, *Ans.*

(3.) $100\% - 40\% = 60\% = \frac{6}{10}$. If $\$3000 = \frac{6}{10}, \frac{1}{10} =$ \$500, and $\frac{4}{10}$ $(40\%) = \$2000$, *Ans.*

(4.) If 56 ct. $= 140\%$ of the cost, the cost $= 56 \div 1.40 = 40$ ct., *Ans.*

(5.) $12\frac{1}{2}\% = \frac{1}{8}$: $\$175 = \frac{7}{8}, \frac{1}{8} = 25$, and $\frac{8}{8} = \$200$, *Ans.*

(6.) $75 \times 4 = 300$: $\frac{1}{8}$ of $300 = 37\frac{1}{2}$: $300 - 37\frac{1}{2} = 262\frac{1}{2}, \times 35$ ct. $= \$91.87\frac{1}{2}$, *Ans.*

(7.) $\$500 - \$425 = \$75$: $7500 \div 500 = 15\%$, *Ans.*

(8.) $100\% - 75\% = 25\%$: $\$5000 = 25\% = \frac{1}{4}$; then $\frac{4}{4} = \$20000$, and $\$20000 - \$5000 = \$15000$, *Ans.*

(9.) $12\frac{1}{2}\% = \frac{1}{8}$: 250 A. 86 sq. rd. $= 40086$ sq. rd. $= \frac{9}{8}$ of neighbor's: $\frac{1}{8} = 4454$, and $\frac{8}{8} = 35632$ sq. rd., $\div 160 = 222$ A. 112 sq. rd., *Ans.*

(10.) $160 \times .35 = 56.00$: $160 + 56 = 216$, *Ans.*

(11.) 5 bu. $\times 32 = 160$ qt.: $6.00 \div 160 = 3\frac{3}{4}\%$, *Ans.*

(12.) $60\% = \frac{6}{10}$: $\frac{6}{10}$ of $45\% = \frac{270}{10} = 27\%$: $540 \div .27 = 2000$ A., *Ans.*

(13.) $371.29 \div 1.07 = \$347$, *Ans.*

(14.) $18 + 15 + 23 + 12 = 68\%$: $100 - 68 = 32\%$: $\$800 \times .32 = \$256$, *Ans.*

(15.) $\frac{1}{20} = 5\%$: $\frac{17}{20} = 17 \times 5 = 85\%$, *Ans.*

(16.) $33\frac{1}{3}\% = \frac{1}{3}$: 2 bu. 3 pk. $= \frac{1}{3}$ of 6 bu. 9 pk. $= 8$ bu. 1 pk., *Ans.*

(17.) 100% less $7\frac{1}{2}\% = 92\frac{1}{2}\%$: $37 \div .925 = 40$, *Ans.*

(18.) $25.8 - 2.58$ $(10\%) = 23.22$ grains, *Ans.*

(19.) $1.25 = \frac{1}{4}$ of 5: $\frac{3}{4}$ remain $= 75\%$, *Ans.*

(20.) $25\% = \frac{1}{4}$. If $\$150 = \frac{5}{4}, \frac{1}{4} = 30$, and $\frac{4}{4} = \$120 =$ cost: $\$200 - \$120 = \$80$: $\frac{80}{120} = \frac{2}{3} = 66\frac{2}{3}\%$, *Ans.*

Art. 172.

(1.) $\$240 \times .05 = \$12$, *Ans.*

(2.) $11.50 \div 460 = .02\frac{1}{2} = 2\frac{1}{2}\%$, *Ans.*

(3.) $\$8.12\frac{1}{2} = 2\frac{1}{2}\%$ of the selling price: $\$8.12\frac{1}{2} \div .02\frac{1}{2} = \$325$, selling price: 1 barrel sold for $\frac{1}{25}$ of $\$325 = \$13$, *Ans.*

(4.) $210 \div 1.05 = \$200$, *Ans.*

(5.) $\$180 \times .04 = \$7.20$: $\$180 - \$7.20 = \$172.80$, *Ans.*

(6.) If $\$11.25 = \frac{1}{20}$ (5%), $\frac{20}{20} = \$225$, *Ans.*

(7.) $\$1323.54 \div 1.08 = \$1225.50$, cost of goods: $\$1323.54 - \$1225.50 = \$98.04$, commission, *Ans.*

(8.)
$$\left.\begin{array}{rcr} 250 \times \$15 & = & \$3750 \\ 175 \times \$7 & = & 1225 \\ 1456 \times \$0.25 & = & 364 \end{array}\right\} = \$5339$$

3% of $\$5339 = \$160.17$: $\$5339 - \$160.17 = \$5178.83$, *Ans.*

Art. 173.

(1.) $20\% = \frac{1}{5}$: $\frac{1}{5}$ of $\$225.50 = \$45.10$: $\$225.50 - \$45.10 = \$180.40$, *Ans.*

(2.) $\frac{1}{3}$ of $\$725.16 = 241.72$: $\$725.16 - 241.72 = 483.44$, $\times .05 = 24.17+$: $\$483.44 - \$24.17 = \$459.27$, *Ans.*

(3.) $100\% - 3\% = 97\%$: $1430.75 \div .97 = \$1475$, *Ans.*

(4.) $100\% - 5\% = 95\%$: $\$390.45 \div .95 = \$411$: $100\% - 25\% = 75\%$: $\$411 \div .75 = \$548$, *Ans.*

(5.) 10% = first discount; $100\% - 10\% = 90\%$: 10% of $90\% = 9\%$, second discount; $90\% - 9\% = 81\%$: 10% of $81\% = 8.1\%$, third discount: $10\% + 9\% + 8.1\% = 27.1\%$, sum of the three discounts: $\$325.20 \div .271 = \$1200$: 1 doz. cost $\frac{1}{20}$ of $\$1200 = \$60$, *Ans.*

(6.) 100 doz. @ 60 ct. $= \$60.00$, less $\$24$ $(40\%) = \$36$, less $\$3.60$ $(10\%) = \$32.40$, less $\$2.43$ $(7\frac{1}{2}\%) = \$29.97$, *Ans.*

(7.) \$50 less 50% = \$25, less 10% = \$22.50, less 10% = \$20.25, less 2% = \$19.845, ÷ 10 = \$1.98+, *Ans.*

Art. 174.

(1.) \$40 + 10% = \$44, *Ans.*

(2.) 5 ct. = $\frac{5}{6}$ the cost; the loss, therefore, is $\frac{1}{6}$ = $16\frac{2}{3}$%, *Ans.*

(3.) $12\frac{1}{2} = \frac{1}{8}$; then 27 ct. = $\frac{9}{8}$ of the cost, $\frac{1}{8}$ = 3 ct., and $\frac{8}{8}$ or the cost = 24 ct., *Ans.*

(4.) \$15.30 ÷ .04 = \$382.50, *Ans.*

(5.) $37\frac{1}{2}\% = \frac{3}{8}$: \$8 + its $\frac{3}{8}$ = \$11, *Ans.*

(6.) 90 — 75 = 15 = $\frac{1}{5}$ of 75 = 20%, *Ans.*

(7.) $6\frac{1}{4}\% = \frac{1}{16}$. If 5 ct. = $\frac{1}{16}$ of the cost, the cost = 80 ct., *Ans.*

(8.) $18\frac{3}{4}\% = \frac{3}{16}$; then \$4.75 = $\frac{19}{16}$, $\frac{1}{16}$ = \$0.25, and $\frac{16}{16}$ = \$0.25 × 16 = \$4, *Ans.*

(9.) \$1.35 = $\frac{9}{10}$ of the cost, $\frac{1}{10}$ = \$0.15, and $\frac{10}{10}$ = \$1.50 = cost: $16\frac{2}{3}\% = \frac{1}{6}$: $\frac{1}{6}$ of 1.50 = .25: \$1.50 + \$0.25 = \$1.75, *Ans.*

(10.) $25\% = \frac{1}{4}$: $\frac{1}{4}$ of \$874 = \$218.50, *Ans.*

(11.) \$1.75 — \$0.25 = \$1.50 : 25 = $\frac{1}{6}$ of 150 = $16\frac{2}{3}$%, *Ans*

(12.) On the first horse \$150 = $\frac{5}{4}$ cost, $\frac{1}{4}$ = \$30, and the cost = \$120: on the second horse \$150 = $\frac{3}{4}$ cost, $\frac{1}{4}$ = \$50 and the cost = \$200: \$200 + \$120 = \$320, — \$300 = \$20, *Ans.*

(13.) 5 ct. = 10% — 8% = 2% of the cost per yd. If 2% = 5 ct., 1% = $2\frac{1}{2}$ ct., and 100% = \$2.50, *Ans.*

(14.) 60 ct. × 10000 = \$6000, cost of the corn: 65 ct. × 7000 = \$4550: 10000 — 7000 = 3000: 55 ct. × 3000 = \$1650: \$4550 + \$1650 = \$6200, the selling price: \$6200 — \$6000 = \$200 gain: \$200 = $\frac{1}{30}$ of \$6000 = $3\frac{1}{3}$%, *Ans.*

(15.) $33\frac{1}{3}\% = \frac{1}{3}$; then $\$12000 = \frac{4}{3}$, $\frac{1}{3} = \$3000$, and $\frac{3}{3} = \$9000 =$ cost of house and lot. The profit was \$3000. On the city lots he lost $\frac{1}{3}$. $\frac{1}{3}$ of $12000 = 4000$: $\$4000 - \$3000 = \$1000$, *Ans.*

Art. 175.

(1.) $100\% - 20\% = 80\%$, cost price. If he sell at the list price, he will gain $\frac{20}{80} = \frac{1}{4} = 25\%$, *Ans.*

(2.) $74 \times 5 \times 45 = 166.50$, less 3.33 (2%) $= \$163.17$: $12\frac{1}{3}\%$ of this amount $= \$20.12+$, *Ans.*

(3.) \$45 less $5\% = \$42.75$, $\div 12 = \$3.56\frac{1}{4} =$ cost per pair: $\$4.25 - \$3.56\frac{1}{4} = \$0.68\frac{3}{4} =$ gain per pair: $5 \times 12 = 60 =$ number of pairs: $60 \times \$0.68\frac{3}{4} = \$41.25$, *Ans.*

(4.) The profit on 36 hats equals 36 times $37\frac{1}{2}$ ct., which is \$13.50. If $\$13.50 = \frac{1}{8}$ of the cost, $\frac{8}{8}$ are 8 times $\$13.50 = \$108$: $\$108 = \frac{9}{10}$ of the list price: $\frac{1}{10} = \$12$, $\frac{10}{10} = \$120$, *Ans.*

(5.) $\$1 \times 100 = \$100$: \$100 less $60\% = \$40$: \$40 less $5\% = \$38$: \$38 less $5\% = \$36.10$: $\$36.10 + \$23.90 = \$60$: $\$60 \div (100 \times 12)$ or $1200 = 5$ ct., *Ans.*

(6.) 100 bbl. @ $\$9.50 = \$950$, less $2\frac{1}{2}\% = \$926.25$, less $\$17.25 = \$909$: $\$909 - (100 \times \$7.50)$ or $\$750 = \$159$, *Ans.*

(7.) $80 \times \$125 = \$10000$, $+ \$200 = \$10200$: $\$10450 - \$10200 = \$250 = 2\frac{1}{2}\%$ of \$10000. *Ans.* $2\frac{1}{2}\%$.

(8.) 1500 lb. $\times 50 = 75000$ lb.: $10\frac{1}{2}$ ct. $\times 75000 = \$7875$: 2% of $\$7875 = \$157.50$, commission: $\$157.50 + \$22.50$, charges, $= \$180$: $\$7875 - \$180 = \$7695$, that the consignor receives: $\$7695 = 114\%$ of the cost price: $\$7695 \div 1.14 = \$6750$, cost: $\$6750 \div 75000 = 9$ ct., cost per pound.

(9.) 60 × 70 ct. = \$42.00: \$42 less 50% and 10% and 5% = \$17.955: \$42 less 20% and 10% and 5% = \$28.728: \$28.728 — \$17.955 = \$10.773, *Ans.*

(10.) \$35.91 = 112% of the cost: \$35.91 ÷ 1.12 = \$32.06¼, the cost: \$32.06¼ ÷ .95 = \$33.75: \$33.75 ÷ .90 = \$37.50: \$37.50 ÷ .75 = \$50 = the list price: \$50 ÷ 50 = \$1, list price per gross, *Ans.*

Art. 177.

(1.) 1% of \$7500 is \$75, and ¼% is ¼ of \$75 = \$18.75, *Ans.*

(2.) 50 shares = \$5000: \$6.25 ÷ \$5000 = .00125 = ⅛%, *Ans.*

(3.) \$10 = ¼% of the investment: 1% = 4 times \$10 = \$40: 100% = 100 times \$40 = \$4000 = 40 shares, *Ans.*

(4.) 1% on \$1700 = \$17.00, and ¼% = \$4.25, *Ans.*

(5.) 95 shares = \$9500: \$11.875 ÷ \$9500 = .00125 = ⅛%, *Ans.*

(6.) If \$9.50 = ¼%, 1% = \$38, and 100% = \$3800 = 38 shares, *Ans.*

Art. 178.

(1.) The dividend will be 3500 times 4 ct. or (\$0.04) = \$140, *Ans.*

(2.) If \$300 is 7½% = $\frac{15}{2}$%, ½% is \$20, and 1% is \$40. If \$40 = 1%, then 100% is \$4000 = 40 shares, *Ans.*

(3.) 15% on \$8000 = \$1200, *Ans.*

(4.) 5% = $\frac{1}{20}$: $\frac{1}{20}$ of 60 = 3: 60 + 3 = 63 shares, *Ans.*

(5.) \$15700 - \$4500 = \$11200: 11200 ÷ 160000 = .07 = 7%, *Ans.*

Art. 179.

(1.) 150 shares of $50 each are equivalent to 75 shares of $100. $139\frac{3}{4} \times 75 = \$10481.25$

$\frac{1}{4}\%$ brokerage on $7500 = 18.75

$10500.00, *Ans.*

(2.) $\$8000 \times 1.10 = \$8800$

$\frac{1}{8}\%$ brokerage on $8000 = 10

$8810, *Ans.*

(3.) $\frac{1}{4}\%$ brokerage on 50 shares = $12.50: $2475 + $12.50 = $2487.50: $2487.50 \div 50 = 49\frac{3}{4}\%$, *Ans.*

(4.) $\$25000 \times 1.14\frac{1}{4} = \$28562.50$

$\frac{1}{8}\%$ brokerage on $25000 = 31.25

$28593.75, *Ans.*

(5.) $19\frac{1}{4} + \frac{1}{4} = 19\frac{1}{2}$ or 19.5: $\$1560 \div 19.5 = 80$, *Ans.*

(6.) $100 \div 1.05 = 95\frac{5}{21}$ ct., *Ans.*

(7.) $1.12\frac{1}{2} = 1\frac{1}{8} = \frac{9}{8}$: $100 \times \frac{8}{9} = \frac{800}{9} = 88\frac{8}{9}$ ct., *Ans.*

(8.) $35\frac{5}{7} = \frac{250}{7}$: $100 \times \frac{7}{250} = 2.80 = 280$, *Ans.*

(9.) $\$8946.25 \div 1.0525 = \$8500$, *Ans.*

(10.) If $\$15.62\frac{1}{2} = \frac{1}{16}\%$, $1\% = \$250$, and $100\% = \$25000$: $\$25734.37\frac{1}{2} - \$25000 = \$734.375$; adding the brokerage to this = $750: $\$750 \div 25000 = .03$ = gold premium: 103 = *Ans.*

Art. 180.

(1.) $\$39900 \times .06 = \$2394$, *Ans.*

(2.) $\$39900 \div 1.05 = \$38000$: $\$38000 \times .06 = \$2280$, *Ans.*

(3.) $\$39900 \div .95 = \$42000$: $42000 \times .06 = \$2520$, *Ans.*

(4.) If gold was *at par*, 6% interest on \$20000 would be \$1200; at 7% premium it would yield an income 7% greater = \$1284, *Ans.*

(5.) \$5220 ÷ 1.16 = \$4500 = amount in bonds; \$4500 × .06 (%) = \$270. Gold being at 5% premium, add to \$270 its 5% = \$13.50; \$270 + \$13.50 = \$283.50, *Ans.*

(6.) $4\frac{1}{2}$ per cents, when gold is at 105, would yield an income $\frac{1}{20}$ greater than when at par. 4.5 + ($\frac{1}{20}$ of 4.5) or .225 = 4.725: 4.725 ÷ 1.08 = $4\frac{3}{8}$%, *Ans.*

(7.)

$$37\tfrac{1}{2}\% = \tfrac{3}{8}:\ \frac{\overset{2}{\cancel{6}}\ \%}{1} \times \frac{8}{\cancel{3}} = 16\%, \textit{ Ans.}$$

(8.) \$1921 ÷ 1.13 = \$1700 = annual income in gold. \$1700 ÷ .05 = \$34000: \$34000 × 1.18 = \$40120, *Ans.*

(9.) $95\frac{1}{4} + \frac{1}{4} = 95\frac{1}{2}$: $105 - \frac{1}{4} = 104\frac{3}{4}$: $104\frac{3}{4} - 95\frac{1}{2} = 9\frac{1}{4}\%$ = .0925: \$925 ÷ .0925 = \$10000 = 100 shares, *Ans.*

(10.) 6 = $\frac{3}{4}$ of 8: $\frac{3}{4}$ = 75%, *Ans.*

(11.) \$4982 ÷ 1.06 = \$4700 = amount of bonds that can be bought: 4% on \$4700 = \$188, *Ans.*

(12.) 7 ÷ $.87\frac{1}{2}$ = .08 = 8%, *Ans.*

(13.) .07 ÷ .06 = $116\frac{2}{3}$, *Ans.*

INTEREST.

Simple Interest.

Art. 183.

1*st. When the time is one year.*

(7.) $6\frac{1}{4}\% = \frac{1}{16}$: \$7200 ÷ 16 = \$450, *Ans.*

(8.) $8\frac{1}{3}\% = \frac{1}{12}$: \$28.20 ÷ 12 = \$2.35: \$28.20 + \$2.35 = \$30.55, *Ans.*

(9.) $10\% = \frac{1}{10}$: $\frac{1}{10}$ of $45.50 = 4.55$: $\$45.50 + \$4.55 = \$50.05$, *Ans.*

(10.) $\$420 \times .05\frac{1}{3} = \$22.40, + \$420 = \$442.40$, *Ans.*

(11.) $\$857 \times .09 = \$77.13, + \$857 = \$934.13$, *Ans.*

(12.) $\$96 \times .08\frac{1}{2} = \$8.16, + \$96 = \$104.16$, *Ans.*

(13.) $\$2000 \times .04\frac{1}{2} = \$90, + \$2000 = \$2090$, *Ans.*

(14.) $12\frac{1}{2}\% = \frac{1}{8}$: $\$164 \div 8 = \$20.50, + \$164 = \$184.50$, *Ans.*

2*d.* *When the time is two or more years.*

(8.) $\$45 \times .08 = \$3.60$: $\$3.60 \times 2 = \$7.20$: $\$7.20 + \$45 = \$52.20$, *Ans.*

(9.) $\$80 \times .07 = \$5.60$: $\$5.60 \times 4 = \$22.40$: $\$80 + \$22.40 = \$102.40$, *Ans.*

(10.) $3\frac{3}{4}\% \times 2 = 7\frac{1}{2}\%$: $\$237.16 \times .07\frac{1}{2} = \$17.79$: $\$237.16 + \$17.79 = \$254.95$, *Ans.*

(11.) $4\% \times 5 = 20\% = \frac{1}{5}$: $\$74.75 \div 5 = \$14.95$: $\$74.75 + \$14.95 = \$89.70$, *Ans.*

(12.) $\$85.45 \times .06 = \$5.127$: $\$5.127 \times 4 = \$20.51$: $\$20.51 + \$85.45 = \$105.96$, *Ans.*

(13.) $\$325 \times .05\frac{2}{5} = \$17.55$: $\$17.55 \times 3 = \$52.65$: $\$52.65 + \$325 = \$377.65$, *Ans.*

(14.) $\$129.36 \times .04\frac{3}{8} = \$5.6595$: $\$5.6595 \times 4 = \$22.638$: $\$22.64 + \$129.36 = \$152$, *Ans.*

(15.) $\$8745 \times .16 = \$1399.20, + \$8745 = \$10144.20$, *Ans.*

3*d.* *When the time is any number of months.*

(2.) $300 @ 6%, 1 yr. $= \$18$: 1 mo. $= \frac{1}{12}$ yr.: $\$18 \div 12 = \$1.50$, *Ans.*

(3.) $\$240 \times .08 = \$19.20$: 2 mo. $= \frac{1}{6}$ yr.: $\$19.20 \div 6 = \$3.20$, *Ans.*

(4.) $\$50 \times .06 = \$3.00$: 4 mo. $= \frac{1}{3}$ yr.: $\frac{1}{3}$ of $\$3.00 = 1.00$
1 mo. $= \frac{1}{4}$ of 4 mo.: $\frac{1}{4}$ of $1.00 = .25$
Interest 5 mo. $= \$1.25$

(5.) $\$86 \times .06 = \$5.16$: 3 mo. $= \frac{1}{4}$ yr.: $\frac{1}{4}$ of $\$5.16 = \$1.29$, *Ans.*

(6.) $\$50 \times .08 = \$4.00$: 4 mo. $= \frac{1}{3}$ yr.: $\frac{1}{3}$ of $\$4.00 = \$1.33+$, *Ans.*

(7.) $\$150.25 \times .08 = \$12.0200$: 6 mo. $= \frac{1}{2}$ yr.: $\frac{1}{2}$ of $\$12.02 = \$6.01$: $\$150.25 + \$6.01 = \$156.26$, *Ans.*

(8.) $\$360 \times .05 = \$18$: 6 mo. $= \frac{1}{2}$ yr.: $\frac{1}{2}$ of $\$18 = \$9$: 1 mo. $= \frac{1}{6}$ of 6 mo.: $\frac{1}{6}$ of $\$9 = \$1.50$: $\$9 + \$1.50 = \$10.50 =$ int. 7 mo.: $\$360 + \$10.50 = \$370.50$, *Ans.*

(9.) $\$204 \times .07 = \$14.28$ 10 mo. $= \frac{10}{12}$ yr.
$\frac{10}{12}$ of $14.28 = 11.90$ 1 mo. $= \frac{1}{10}$ of 10 mo.
$\frac{1}{10}$ of $11.90 = 1.19$
Int. 11 mo. $= \$13.09, + \$204 = \$217.09$. *Ans.*

(10.) $\$228 \times .06 = \$13.68$ 6 mo. $= \frac{1}{2}$ yr.
$\frac{1}{2}$ of $13.68 = 6.84$ 3 mo. $= \frac{1}{2}$ of 6 mo.
$\frac{1}{2}$ of $6.84 = 3.42$
Int. 9 mo. $= \$10.26, + \$228 = \$238.26$, *Ans.*

(11.) $\$137.50 \times .06 = \$8.25$: 8 mo. $= \frac{2}{3}$ yr.: $\frac{1}{3}$ of $\$8.25 = \$2.75$, $\frac{2}{3} = \$5.50$: $\$137.50 + \$5.50 = \$143$, *Ans.*

(12.) $\$7596 \times .08 = \$607.68$: 10 mo. $= \frac{5}{6}$ yr.: $\frac{5}{6}$ of $\$607.68 = \$506.40$: $\$7596 + \$506.40 = \$8102.40$, *Ans.*

4th. When the time is any number of days.

(2.) $\$360 \times .06 = 21.60$: $\frac{1}{12}$ (1 mo.) of $\$21.60 = \$1.80$: 20 da. $= \frac{2}{3}$ mo.: $\frac{2}{3}$ of $\$1.80 = \$1.20$, *Ans.*

(3.) $726 × .06 = $43.56: $\frac{1}{12}$ (1 mo.) of $43.56 = $3.63: 10 da. = $\frac{1}{3}$ mo.: $\frac{1}{3}$ of $3.63 = $1.21, *Ans.*

(4.) $1200 × .06 = $72: $\frac{1}{12}$ of $72 = $6 = int. 1 mo.: 15 da. = $\frac{1}{2}$ mo.: $\frac{1}{2}$ of $6 = $3, *Ans.*

(5.) $180 × .08 = $14.40: $\frac{1}{12}$ of $14.40 = $1.20 (1 mo.).

15 da. = $\frac{1}{2}$ mo. = $0.60
3 da. = $\frac{1}{5}$ 15 da. = .12
1 da. = $\frac{1}{3}$ 3 da. = .04
Int. for 19 da. = $0.76, *Ans.*

(6.) $240 × .07, ÷ 12 = $1.40 = int. 1 mo.

24 da. = $\frac{4}{5}$ mo. = $1.12
3 da. = $\frac{1}{8}$ 24 da. = .14
Int. 27 da. = $1.26, *Ans.*

(7.) $320 × .05, ÷ 12 = 1.33\frac{1}{3}$ = int. 1 mo.

20 da. = $\frac{2}{3}$ mo. = $0.888+
1 da. = $\frac{1}{20}$ 20 da. = .044+
Int. 21 da. = $0.93, *Ans.*

(8.) $450 × .10, ÷ 12 = $3.75 = int. 1 mo.: 25 da. = $\frac{5}{6}$ mo.: $\frac{5}{6}$ of $3.75 = $3.125. *Ans.* $3.13.

(9.) $100.80 × .05, ÷ 12 = $0.42 = int. 1 mo.

25 da. = $\frac{5}{6}$ mo. = $0.35
3 da. = $\frac{1}{10}$ mo. = .042
Int. 28 da. = $0.39, + $100.80 = $101.19, *Ans.*

(10.) $150 × .05, ÷ 12 = 0.62\frac{1}{2}$: 18 da. = $\frac{3}{5}$ mo.: $\frac{3}{5}$ of $0.625 = $0.375: $150 + $0.375 = $150.38, *Ans.*

(11.) $\$360 \times .06, \div 12 = \$1.80 =$ int. 1 mo.

10 da. $= \frac{1}{3}$ mo. $= \$0.60$

1 da. $= \frac{1}{10}$ 10 da. $= \underline{\quad .06}$

Int. for 11 da. $= \$0.66, + \$360 = \$360.66$, *Ans.*

(12.) $\$264 \times .06, \div 12 = \$1.32 =$ int. 1 mo.

6 da. $= \frac{1}{5}$ mo. $= \$0.264$

3 da. $= \frac{1}{2}$ 6 da. $= \underline{\quad .132}$

Int. 9 da. $= \$0.40, + \$264 = \$264.40$, *Ans.*

(13.) $\$900 \times .07 = \$63, \div 12 = \$5.25 =$ int. 1 mo.

10 da. $= \frac{1}{3}$ mo. $= \$1.75$

3 da. $= \frac{1}{10}$ mo. $= \quad .525$

1 da. $= \frac{1}{3}$ 3 da. $= \underline{\quad .175}$

Int. 14 da. $= \$2.45, + \$900 = \$902.45$, *Ans.*

(14.) $\$430 \times .04\frac{1}{2}, \div 12 = \$1.61\frac{1}{4} =$ int. 1 mo.

15 da. $= \frac{1}{2}$ mo. $= \$0.806$

3 da. $= \frac{1}{5}$ 15 da. $= \quad .161$

1 da. $= \frac{1}{3}$ 3 da. $= \underline{\quad .053}$

Int. 19 da. $= \$1.02, + \$430 = \$431.02$, *Ans.*

5*th. When the time is years, months, and days, or any two of these periods.*

(3.) $\$150 \times .06 = \$9, \times 4$ (yr.) $= \$36.00$: 2 mo. $= \frac{1}{6}$ yr.: $\frac{1}{6}$ of $\$9 = \$1.50$: $\$36 + \$1.50 = \$37.50$, *Ans.*

(4.) $\$375.40 \times .06 = \$22.524$ 8 mos. $= \frac{2}{3}$ yr.

$\frac{2}{3}$ of $\$22.524 = \underline{\quad 15.016}$

$\$37.54$, *Ans.*

(5.) $\$92.75 \times .06 = \$5.565, \times 3$ (yr.) $= \$16.695$

4 mo. $= \frac{1}{3}$ yr. $\frac{1}{3}$ of $5.565 = \quad 1.855$

1 mo. $= \frac{1}{4}$ 4 mo. $\frac{1}{4}$ of $1.855 = \underline{\quad .46375}$

$\$19.01$, *Ans.*

(6.) $\$500 \times .06 = \$30.00 =$ int. 1 yr.
$\frac{1}{12}$ of $\$30.00 = 2.50 =$ int. 1 mo.
$\frac{1}{2}$ of $2.50 = 1.25 =$ int. 15 da.
$\frac{1}{5}$ of $1.25 = .25 =$ int. 3 da.
$34.00, *Ans.*

(7) $\$560 \times .08 = \$44.80, \times 2 = \$89.60 =$ int. 2 yr.
$\frac{1}{3}$ of $44.80 = 14.933+ =$ int. 4 mo.
$\frac{1}{8}$ of $14.933 = 1.866 =$ int. 15 da.
$106.40, *Ans.*

(8.) $\$750 \times .06 = \$45, \times 4 = \$180.00 =$ int. 4 yr.
3 mo. $= \frac{1}{4}$ yr. $\frac{1}{4}$ of $\$45.00 = 11.25 =$ int. 3 mo.
Int. 1 mo. $= \frac{1}{3}$ of $11.25 = 3.75$:
$\frac{1}{5}$ of $3.75 = .75 =$ int. 6 da.
$192, *Ans.*

(9.) $\$456 \times .05 = \$22.80, \times 3 = \$68.40 =$ int. 3 yr.
$\frac{1}{12}$ of $\$22.80 = \$1.90$, $\frac{5}{12} = 9.50 =$ int. 5 mo.
$\frac{1}{2}$ of $1.90 = .95 =$ int. 15 da.
$\frac{1}{5}$ of $.95 = .19 =$ int. 3 da.
$79.04, *Ans.*

(10.) $\$216 \times .10 = \$21.60, \times 5 = \$108.00 =$ int. 5 yr.
$\frac{1}{12}$ of $\$21.60 = 1.80$ (1 mo.),
$\frac{7}{12}$ of $21.60 = 12.60 =$ int. 7 mo.
27 da. = 3 da., or $\frac{1}{10}$ less than 1
mo.: $\frac{1}{10}$ of $1.80 = .18$: $1.80 — .18 = 1.62 =$ int. 27 da.
$122.22, *Ans.*

(11.) $\$380 \times .15 = \$57.00, \times 3 = \$171.00 =$ int. 3 yr.
9 mo. $= \frac{3}{4}$ yr.: $\frac{3}{4}$ of $\$57 = 42.75 =$ int. 9 mo.
$\frac{1}{12}$ of $\$57 = \$4.75$ (1 mo.): 9 da.
$= \frac{3}{10}$ mo.: $\frac{3}{10}$ of $\$4.75 = 1.425 =$ int. 9 da.
$215.18, *Ans.*

(12.) $300 × .06 = $18, × 3 = $54.00 = int. 3 yr.
$\frac{1}{2}$ of $18 = 9.00 = int. 6 mo.
$\frac{1}{3}$ of 9 = 3.00 = int. 2 mo.
Int. 3 yr. 8 mo. = $66.
300.
$366, *Ans.*

(13.) $250 × .06 = $15.00 = int. 1 yr.
$\frac{1}{2}$ of $15.00 = 7.50 = int. 6 mo.
$\frac{1}{6}$ of 7.50 = 1.25 = int. 1 mo.
$23.75
250.00
$273.75, *Ans.*

(14.) $205.25 × .06 = $12.315, × 2 = $24.63 (2 yr.)
Int. 6 mo. = $\frac{1}{2}$ of $12.315 = 6.1575
Int. 2 mo. = $\frac{1}{3}$ of 6.1575 = 2.0525
Int. 15 da. = $\frac{1}{4}$ of 2.0525 = .5131
$33.3531
205.25
$238.60, *Ans.*

(15.) $150.62 × .05 = $7.5310, × 3 = $22.5930 (3 yr.)
Int. 4 mo. = $\frac{1}{3}$ of $7.5310 = 2.5103
Int. 1 mo. = $\frac{1}{4}$ of 2.5103 = .6275
Int. 12 da. = $\frac{2}{5}$ of .6275 = .2510
$25.9818
150.62
$176.60, *Ans.*

(16.) $210.25 × .07 = $14.7175, × 2 = $29.4350 (2 yr.)
Int. 6 mo. = $\frac{1}{2}$ of $14.7175 = 7.3587
Int. 1 mo. = $\frac{1}{6}$ of 7.3587 = 1.2264
Int. 20 da. = $\frac{2}{3}$ of 1.2264 = .8176
$38.8377
210.25
$249.09, *Ans.*

(17.) $57.85 × .05 = $2.8925, × 2 = $5.7850 (2 yr.)
Int. 3 mo. = $\frac{1}{4}$ of $2.8925 = .7231
Int. 20 da. = $\frac{2}{3}$ of $\frac{1}{3}$ of .7231 = .1606
Int. 2 da. = $\frac{1}{10}$ of .1606 = .0160
Int. 1 da. = $\frac{1}{2}$ of .0160 = .0080
$6.6927
57.85
$64.54, *Ans.*

(18.)

yr.	mo.	da.
1849	4	19
1847	1	9
2	3	10

$150 × .06 = $9.00, × 2 = $18.00 = int. 2 yr.
Int. 3 mo. = $\frac{1}{4}$ of $9.00 = 2.25
Int. 10 da. = $\frac{1}{3}$ of $\frac{1}{3}$ of 2.25 = .25
$20.50, *Ans.*

(19.)

yr.	mo.	da.
1849	4	27
1848	2	15
1	2	12

$240 × .08 = $19.20 = int. 1 yr.
$\frac{1}{6}$ of $19.20 = 3.20 = int. 2 mo.
$\frac{2}{5}$ of $\frac{1}{2}$ of 3.20 = .64 = int. 12 da.
$23.04, *Ans.*

(20.)

yr.	mo.	da.
1845	8	28
1843	5	14
2	3	14

$\$180 \times .07 = \$12.60, \times 2 = \$25.20 = \text{int. } 2 \text{ yr.}$
$\frac{1}{4}$ of $12.60 = 3.15 = int. 3 mo.
$\frac{2}{5}$ of $\frac{1}{3}$ of 3.15 = .42 = int. 12 da.
$\frac{1}{6}$ of .42 = .07 = int. 2 da.
$28.84, *Ans.*

(21.)

mo.	da.
11	27
7	3
	24

$\$137.50 \times .09 = \$12.3750 = \text{int. } 1 \text{ yr.}$
$\frac{1}{3}$ of $12.3750 = $4.125 = int. 4 mo.
$\frac{4}{5}$ of $\frac{1}{4}$ (= $\frac{1}{5}$) of 4.125 = .825 = int. 24 da.
$4.95, *Ans.*

(22.)

mo.	da.
8	28
3	1
5	27

$\$125.40 \times .08\frac{1}{2} = \$10.659 = \text{int. } 1 \text{ yr.}$
$\frac{5}{12}$ of $10.659 = $4.44 + = int. 5 mo.
$\frac{9}{10}$ of $\frac{1}{5}$ of 4.44 = .799+ = int. 27 da.
$5.24
125.40
$130.64, *Ans.*

(23.)

yr.	mo.	da.
1848	3	9
1847	8	2
	7	7

$\$234.60 \times .05\frac{1}{4} = \$12.3165 = \text{int. 1 yr.}$

$\frac{7}{12}$ of \$12.3165 = \$7.1848 = int. 7 mo.
$\frac{1}{5}$ of $\frac{1}{12}$ of 12.3165 = .2052 = int. 6 da.
$\frac{1}{6}$ of .2052 = .0342 = int. 1 da.

\$7.4242
234.60
\$242.02, *Ans.*

(24.)

yr.	mo.	da.
1847	7	24
1846	10	25
	8	29

$\$153.80 \times .05 = \$7.69 = \text{int. 1 yr.}$

$\frac{2}{3}$ of \$7.69 = \$5.126 = int. 8 mo.
Int. 1 mo. = $\frac{1}{8}$ of \$5.126 = .64
Int. 29 da. = .64 less $\frac{1}{30}$ = .62

\$5.75
153.80
\$159.55, *Ans.*

Art. 184. 1st Process.

(5.) 1 yr. 4 mo. = 16 mo. *Ans.* 16 ct.

(6.) 1 yr. 5 mo. = 17 mo.: $\frac{1}{3}$ of 27 da. = 9. *Ans.* \$0.179

(7.) 2 yr. 3 mo. = 27 mo.: $\frac{1}{3}$ of 21 da. = 7. *Ans.* \$0.277

(8.) 3 yr. 7 mo. = 43 mo.: $\frac{1}{3}$ of 12 da. = 4. *Ans.* \$0.434

(9.) 4 yr. 2 mo. = 50 mo.: $\frac{1}{3}$ of 15 da. = 5 da. *Ans.* \$0.505

(10.) 2 ct. for the 2 mo., and $\frac{1}{3}$ mill for the 1 da. *Ans.* \$0.020$\frac{1}{3}$

(11.) $\frac{1}{3}$ of 17 = 5$\frac{2}{3}$ (apply rule). *Ans.* \$0.055$\frac{2}{3}$

(12.) $\frac{1}{3}$ of 13 = 4$\frac{1}{3}$ (apply rule). *Ans.* \$0.104$\frac{1}{3}$

(13.) 1 yr. 2 mo. = 14 mo.: $\frac{1}{3}$ of 4 = 1$\frac{1}{3}$. *Ans.* \$0.141$\frac{1}{3}$

(14.) 2 yr. 9 mo. = 33 mo.: $\frac{1}{3}$ of 20 = 6$\frac{2}{3}$. *Ans.* \$0.336$\frac{2}{3}$

(15.) 3 yr. 5 mo. = 41 mo.: $\frac{1}{3}$ of 29 = 9$\frac{2}{3}$. *Ans.* \$0.419$\frac{2}{3}$

2D PROCESS.

(3.) Int. on \$1 for 7 mo. 24 da. @ 12% = \$0.078: int. for 7 mo. 24 da. @ 6% = $\frac{1}{2}$ of \$0.078 = \$0.039, *Ans.*

(4.) Int. at 12% = \$0.105

Int. at 4% = $\frac{1}{3}$ of \$0.105 = \$0.035
Int. at 1% = $\frac{1}{4}$ of .035 = .00875

\$0.043$\frac{3}{4}$, *Ans.*

(5.) Int. at 12% = \$0.116: int. at 9% = $\frac{3}{4}$ of \$0.116 = \$0.087, *Ans.*

(6.) 1 yr. 2 mo. = 14 mo.: int. 14 mo. 9 da. @ 12% = \$0.143: int. @ 6% = $\frac{1}{2}$ of \$0.143 = \$0.071$\frac{1}{2}$, *Ans.*

(7.) 2 yr. 5 mo. = 29 mo.: int. 29 mo. 12 da. @ 12% = \$0.294: int. @ 8% = $\frac{2}{3}$ of \$0.294 = \$0.196, *Ans.*

(8.) 3 yr. 10 mo. = 46 mo.: int. 46 mo. 17 da. @ 12% = \$0.465$\frac{2}{3}$: int. @ 10% = $\frac{5}{6}$ of \$0.465$\frac{2}{3}$ = \$0.388$\frac{1}{18}$, *Ans.*

(9.) 4 yr. 3 mo. = 51 mo.: int. 51 mo. 11 da. @ 12% = \$0.513$\frac{2}{3}$

Int. @ 6% = $\frac{1}{2}$ of \$0.513$\frac{2}{3}$ = \$0.256$\frac{5}{6}$
Int. @ 1% = $\frac{1}{6}$ of .256$\frac{5}{6}$ = .042$\frac{29}{36}$

\$0.299$\frac{23}{36}$, *Ans.*

(10.) 5 yr. 7 mo. = 67 mo.: int. 67 mo. 24 da. @ 12% = \$0.678: int. @ 4% = $\frac{1}{3}$ of \$0.678 = \$0.225, *Ans.*

3D PROCESS.

(3.) Int. \$1 for 6 mo. 21 da. @ 12% = \$0.067: @ 6% = $\frac{1}{2}$ of \$0.067 = \$0.0335, × 40 = \$1.34, *Ans.*

(4.) Int. \$1 for 8 mo. 24 da. @ 12% = \$0.088: int. \$1 for 8 mo. 24 da. @ 9% = $\frac{3}{4}$ of \$0.088 = \$0.066: \$0.066 × 50 = \$3.30, *Ans.*

(5.) Int. \$1 for 10 mo. 12 da. @ 12% = \$0.104

Int. \$1 for 10 mo. 12 da. @ 6% = $\frac{1}{2}$ of \$0.104 = \$0.052
Int. \$1 for 10 mo. 12 da. @ 1% = $\frac{1}{6}$ of .052 = .008$\frac{2}{3}$
\$0.060$\frac{2}{3}$ × 120 = \$7.28, *Ans.* \$0.060$\frac{2}{3}$

(6.) Int. \$1 for 11 mo. 15 da. @ 12% = \$0.115: int. \$1 for 11 mo. 15 da. @ 6% = $\frac{1}{2}$ of \$0.115 = \$0.0575: \$0.0575 × 200 = \$11.50, *Ans.*

(7.) 1 yr. 3 mo. = 15 mo.: int. \$1 for 15 mo. 6 da. @ 12% = \$0.152: at 3% = $\frac{1}{4}$ of \$0.152 = \$0.038: \$0.038 × 500 = \$19, *Ans.*

(8.) 1 yr. 5 mo. = 17 mo.: int. \$1 for 17 mo. 27 da. @ 12% = \$0.179: at 8% = $\frac{2}{3}$ of \$0.179 = \$0.119$\frac{1}{3}$: \$0.119$\frac{1}{3}$ × 750 = \$89.50, *Ans.*

(9.) 1 yr. 9 mo. = 21 mo.: int. \$1 for 21 mo. 3 da. @ 12% = \$0.211: @ 6% = $\frac{1}{2}$ of \$0.211 = \$0.105$\frac{1}{2}$, × 48.75 = \$5.14, *Ans.*

(10.) 1 yr. 10 mo. = 22 mo.: Int. \$1 for 22 mo. 25 da. @ 12% = \$0.228$\frac{1}{3}$: at 4% = $\frac{1}{3}$ of \$0.228$\frac{1}{3}$ = \$0.076$\frac{1}{9}$: \$0.076$\frac{1}{9}$ × 76.32 = \$5.81, *Ans.*

(11.) 2 yr. 1 mo. = 25 mo.: int. \$1 for 25 mo. 9 da. @ 12% = \$0.253: at 4% = $\frac{1}{3}$ of \$0.253 = \$0.084$\frac{1}{3}$: 1% = $\frac{1}{4}$

of 0.084\frac{1}{3}$ = 0.021\frac{1}{12}$: 0.084\frac{1}{3}$ + 0.021\frac{1}{12}$ = 0.105\frac{5}{12}$, × 600 = $63.25: $600 + $63.25 = $663.25, *Ans.*

(12.) 2 yr. 4 mo. = 28 mo.: int. $1 @ 12% for 28 mo. 10 da. = 0.283\frac{1}{3}$: @ 6% = $\frac{1}{2}$ of 0.283\frac{1}{3}$ = 0.141\frac{2}{3}$, × 900 = $127.50: $900 + $127.50 = $1027.50, *Ans.*

(13.) 2 yr. 7 mo. = 31 mo.: int. $1 @ 12% for 31 mo. 17 da. = 0.315\frac{2}{3}$: 9% = $\frac{3}{4}$ of 12% = 0.236\frac{3}{4}$: 0.236\frac{3}{4}$ × 86.25 = $20.419+: $86.25 + $20.42 = $106.67, *Ans.*

(14.) 3 yr. 2 mo. = 38 mo.: int. $1 for 38 mo. 13 da. @ 12% = 0.384\frac{1}{3}$: 8% = $\frac{2}{3}$ of 12% = 0.256\frac{2}{9}$, × 450 = $115.30: $450 + $115.30 = $565.30, *Ans.*

(15.) 3 yr. 5 mo. = 41 mo.: int. $1 for 41 mo. 22 da. @ 12% = 0.417\frac{1}{3}$: 4% = $\frac{1}{3}$ of 12% = 0.139\frac{1}{9}$: 0.139\frac{1}{9}$ × 534.78 = $74.39+: $534.78 + $74.39 = $609.17, *Ans.*

(16.) 3 yr. 11 mo. = 47 mo.: int. $1 @ 12% for 47 mo. 15 da. = $0.475: int. @ 10% = $\frac{5}{6}$ of 12% = 0.395\frac{5}{6}$: 0.395\frac{5}{6}$ × 1200 = $475, + $1200 = $1675, *Ans.*

Art. 185.

(4.) Int. on $200 for 1 yr. @ 6% = $12.00: 36 ÷ 12 = 3. *Ans.* 3 yr.

(5.) Int. on $60 for 1 yr. @ 5% = $3.00: $72 — $60 = $12: 12 ÷ 3 = 4. *Ans.* 4 yr.

(6.) If the principal is doubled, the int. will equal 100%. 100% ÷ 6% = 16$\frac{2}{3}$: $\frac{2}{3}$ yr. = 8 mo. *Ans.* 16 yr 8 mo.

(7.) Int. on $375 for 1 yr. @ 8% = $30: 90 ÷ 30 = 3. *Ans.* 3 yr.

(8.) Int. on $600 @ 9% for 1 yr. = $54: $798 — $600 = $198: 198 ÷ 54 = 3$\frac{2}{3}$ = 3 yr. 8 mo., *Ans.*

(9.) $100(\%) \div 10(\%) = 10$. *Ans.* 10 yr.

(10.) Int. on \$250 for 1 yr. @ 6% = \$15: $34.50 \div 15 = 2.30$ or $2\frac{3}{10}$ yr.: $\frac{3}{10}$ yr. $= 3\frac{3}{5}$ mo.: $\frac{3}{5}$ mo. = 18 da. *Ans.* 2 yr. 3 mo. 18 da.

(11.) The int. on \$60 for 1 yr. @ 6% = \$3.60: \$73.77 — \$60 = \$13.77: $13.77 \div 3.60 = 3.825$ or $3\frac{33}{40}$ yr.: $\frac{33}{40}$ yr. $= 9\frac{9}{10}$ mo.: $\frac{9}{10}$ mo. = 27 da. *Ans.* 3 yr. 9 mo. 27 da.

(12.) If the principal is trebled, the int. will equal 200%. $200(\%) \div 6(\%) = 33\frac{1}{3}$: $\frac{1}{3}$ yr. = 4 mo. *Ans.* 33 yr. 4 mo.

(13.) Int. on \$400 for 1 yr. @ 7% = \$28: $68.60 \div 28 = 2.45$ or $2\frac{9}{20}$ yr.: $\frac{9}{20}$ yr. $= 5\frac{4}{10}$ mo.: $\frac{4}{10}$ mo. = 12 da. *Ans.* 2 yr. 5 mo. 12 da.

(14.) Int. on \$700 for 1 yr. @ 9% = \$63: \$924.70 — \$700 = \$224.70: $224.70 \div 63 = 3.566+$ or $3\frac{57}{100}$ yr.: $\frac{57}{100}$ yr. $= 6\frac{79}{100}$ mo.: $\frac{79}{100}$ mo. $= 23\frac{9}{10}$ da. *Ans.* 3 yr. 6 mo. 24 da.

(15.) If the principal is increased one half, the int. will equal 50%. $50(\%) \div 8(\%) = 6\frac{1}{4}$: $\frac{1}{4}$ yr. = 3 mo. *Ans.* 6 yr. 3 mo.

(16.) Int. on \$1200 for 1 yr. @ 10% = \$120: \$1675 — \$1200 = \$475: $475 \div 120 = 3.959+$ or $3\frac{96}{100}$ yr.: $\frac{96}{100}$ yr. $= 11\frac{52}{100}$ mo.: $\frac{52}{100}$ mo. = 15+ da. *Ans.* 3 yr. 11 mo. 15 da.

Art. 186.

(3.) \$48 ÷ 2 = \$24 = int. 1 yr.: $24 \div 600 = .04 = 4\%$, *Ans.*

(4.) 2 yr. 6 mo. $= 2\frac{1}{2}$ or $\frac{5}{2}$ yr. If int. for $\frac{5}{2}$ yr. = \$200, for $\frac{1}{2}$ yr. = \$40, and for 1 yr. = \$80: $80 \div 1000 = .08 = 8\%$, *Ans.*

(5.) 2 yr. 4 mo. 24 da. $= 2\frac{2}{5}$ or $\frac{12}{5}$ yr.: \$310 — \$250 = \$60. If int. for $\frac{12}{5}$ yr. = \$60, for $\frac{1}{5}$ yr. = \$5, and for 1 yr. = \$25: $25 \div 250 = .10 = 10\%$, *Ans.*

(6.) $\$23.40 \div 2 = \$11.70$ = int. 1 yr.: $11.70 \div 260 = .04\frac{1}{2} = 4\frac{1}{2}\%$, *Ans.*

(7.) Since the int. for $12\frac{1}{2}$ or $\frac{25}{2}$ yr. is 100%, for $\frac{1}{2}$ yr. it is $\frac{100}{25}$ or 4 and for 1 yr. = 8%, *Ans.*

(8.) \$250.25 — \$175 = \$75.25: 3 yr. 7 mo. $= 3\frac{7}{12}$ or $\frac{43}{12}$ yr. Since \$75.25 = int. for $\frac{43}{12}$ yr., for $\frac{1}{12}$ yr. = \$75.25 $\div$ 43 = \$1;75, and for 1 yr. = \$1.75 $\times$ 12 = \$21: $21 \div 175 = .12 = 12\%$, *Ans.*

(9.) 1 yr. 8 mo. 12 da. $= 1\frac{7}{10}$ or $\frac{17}{10}$ yr.: \$61.20 $\div$ 17 = \$3.60, $\times$ 10 = \$36 = int. 1 yr.: $36 \div 450 = 0.08 = 8\%$, *Ans.*

(10.) 11 yr. 1 mo. 10 da. $= 11\frac{1}{9}$ or $\frac{100}{9}$ yr. Since the int. for $\frac{100}{9}$ yr. = 100%, for $\frac{1}{9}$ yr. = 1%, and for 1 yr. = 9%, *Ans.*

(11.) \$746.20 — \$650 = \$96.20: 2 yr. 5 mo. 18 da. $= 2\frac{7}{15}$ or $\frac{37}{15}$ yr.: \$96.20 $\div$ 37 = \$2.60, $\times$ 15 = \$39 = int. 1 yr.: $39 \div 650 = .06 = 6\%$, *Ans.*

(12.) $\$110.40 \div 6 = \$18.40$ = int. 1 yr.: $18.40 \div 640 = .02\frac{7}{8} = 2\frac{7}{8}\%$, *Ans.*

Art. 187.

(3.) The int. of \$1 for 3 yr. at 5% is 15 ct. It will take as many dollars to gain \$8.25 int. as 15 ct. are contained times in \$8.25 = 55 times. *Ans.* \$55.

(4.) Int. of \$1 for 3 yr. at 5% = 15 ct.: \$341.25 $\div$ .15 = \$2275, *Ans.*

(5.) 1 yr. 4 mo. $= 1\frac{1}{3}$ yr.: 6% for 1 yr. = .06, and for $1\frac{1}{3}$ yr. = .08: \$226 $\div$ .08 = \$28.25, *Ans.*

(6.) Int. of \$1 = 5 ct.: \$1023.75 ÷ .05 = \$20475, *Ans.*

(7.) The int. of \$1 for 1 yr. 6 mo. 27 da. at 12% = \$0.189: at 8% the int. is $\frac{2}{3}$ of \$0.189 = \$0.126: \$30.24 ÷ .126 = \$240, *Ans.*

(8.) Int. of \$1 for 12 yr. 3 mo. 20 da. at 12% = \$1.476$\frac{2}{3}$: at 9% = $\frac{3}{4}$ of \$1.476$\frac{2}{3}$ = \$1.1075: \$525.40 ÷ 1.1075 = \$474.40, *Ans.*

(9.) Int. at 12% on \$1 for 2 yr. 7 mo. 11 da. is \$0.313$\frac{2}{3}$: at 4% it is $\frac{1}{3}$ of \$0.313$\frac{2}{3}$ = \$0.104$\frac{5}{9}$: \$9.41 ÷ .104$\frac{5}{9}$ = \$90, *Ans.*

(10.) The int. of \$1 for 5 yr. 8 mo. 24 da. at 12% is \$0.688: at 6% it is $\frac{1}{2}$ of \$0.688 = \$0.344: \$28.38 ÷ .344 = \$82.50, *Ans.*

Art. 188.

(2.) 9 yr. × .05 = .45: \$435 ÷ 1.45 = \$300, *Ans.*

(3.) 4 yr. × .05 = .20: \$571.20 ÷ 1.20 = \$476 = principal: \$571.20 — \$476 = \$95.20, *Ans.*

(4.) 6 yr. × .07 = 0.42: \$532.50 ÷ 1.42 = \$375: \$532.50 — \$375 = \$157.50, *Ans.*

(5.) 2 yr. 9 mo. = 2$\frac{3}{4}$ yr.: 2$\frac{3}{4}$ × .08 = 0.22: \$285.48 ÷ 1.22 = \$234, *Ans.*

(6.) 2$\frac{1}{2}$ yr. × .06 = 0.15: \$690 ÷ 1.15 = \$600: \$690 — \$600 = \$90, *Ans.*

(7.) 3 yr. 4 mo. 24 da. = 3$\frac{2}{5}$ yr.: 3$\frac{2}{5}$ × .07 = 0.238: \$643.760 ÷ 1.238 = \$520, *Ans.*

(8.) 4 yr. 3 mo. 27 da. = $\frac{519}{120}$ yr.: $\frac{519}{120}$ × .04 = 0.173: \$914.940 ÷ 1.173 = \$780 = principal: \$914.94 — \$780 = \$134.94, *Ans.*

COMPOUND INTEREST.

Art. 190.

(2)

$500	530	561.80
.06	.06	.06
30.00	31.80	33.7080
500	530	561.80
$530, 1st yr.	$561.80, 2d yr.	*Ans.* $595.51

(3)

$800	848	898.88	952.81
.06	.06	.06	.06
48.00	50.88	53.9328	57.1686
800	848	898.88	952.81
$848, 1st yr.	$898.88, 2d yr.	$952.81, 3d yr.	$1009.98, *Ans.*

(4)

$250	15.90
.06	265
$15.00 = 1st yr.	$280.90
250	.06
265	$16.8540 = 3d yr.
.06	
$15.90 = 2d yr.	$15+$15.90+$16.85=$47.75, *Ans.*

(5)

$300	330.75
.05	.05
$15.00 = 1st yr.	$16.5375 = 3d yr.
300	330.75
315	347.29
.05	.05
$15.75 = 2d yr.	$17.3645 = 4th yr.
315	
$330.75	$15+$15.75+$16.54+$17.36=$64.65, *Ans.*

(5)

$200
.03
$6.00 = 1st hf.-yr.
200
206
.03
$6.18 = 2d hf.-yr.
206
$212.18

212.18
.03
$6.3654 = 3d hf.-yr.
212.18
218.55
.03
$6.5565 = 4th hf.-yr.

$6 + $6.18 + $6.36 + $6.56 = $25.10, *Ans.*

(7.) 20% annually = 5% quarterly.
1st qr., $500 × .05 = $25, + $500 = $525:
2d qr., $525 × .05 = $26.25, + $525 = $551.25:
3d qr., $551.25 × .05 = $27.56, + $551.25 = $578.81:
4th qr., $578.81 × .05 = $28.94, + $578.81 = $607.75:
5th qr., $607.75 × .05 = $30.39, + $607.75 = $638.14:
6th qr., $638.14 × .05 = $31.91, + $638.14 = $670.05:
7th qr., $670.05 × .05 = $33.50, + $670.05 = $703.55:
8th qr., $703.55 × .05 = $35.18, + $703.55 = $738.73, *Ans.*

(8.) Int. on $300, 1 yr. @ 6% = $18, + $300 = $318: int. on $318 for 1 yr. @ 6% = $19.08, + $318 = $337.08: int. on $337.08, ½ yr. @ 6% = $10.11, + $337.08 = $347.19: $347.19 — $300 = $47.19, *Ans.*

(9.) Int. on $1000, 1 yr. @ 6% = $60, + $1000 = $1060: int. on $1060, 1 yr. @ 6% = $63.60, + $1060 = $1123.60: int. on $1123.60 for 8½ mo. @ 6% = $47.75, + $1123.60 = $1171.35: $1171.35 — $1000 = $171.35, *Ans.*

(10.) 6% int. annually = 3% semi-annually.
Int. 6 mo. on $620 @ 3% = $18.60, + $620 = $638.60:
int. 6 mo. on $638.60 @ 3% = $19.16, + $638.60 = $657.76:
int. 6 mo. on $657.76 @ 3% = $19.73, + $657.76 = $677.49:
int. 6 mo. on $677.49 @ 3% = $20.32, + $677.49 = $697.81:

int. 6 mo. on \$697.81 @ 3% = \$20.93, + \$697.81 = \$718 74: int. 6 mo. on \$718.74 @ 3% = \$21.56, + \$718.74 = \$740.31: int. 6 mo. on \$740.31 @ 3% = \$22.21, + \$740.30 = \$762.52, *Ans.*

(1st.—Compound Interest.)

(11.) 1st yr., int. on \$500 @ 6% = \$30, + \$500 = \$530: 2d yr., int. on \$530 @ 6% = \$31.80, + \$530 = \$561.80: 3d yr., int. on \$561.80 @ 6% = \$33.71, + \$561.80 = \$595.51: 4th yr., int. on \$595.51 @ 6% = \$35.73, + \$595.51 = \$631.24: 8 mo. = $\frac{2}{3}$ yr., int. on \$631.24 @ 6% = \$25.25, + \$631.24 = \$656.49: \$656.49 — \$500 = \$156.49

(2d.—Simple Interest.)

Int. on \$500 for 1 yr. @ 6% = \$30: int. on \$500 for $4\frac{2}{3}$ yr. = \$140. \$156.49 — \$140 = \$16.49, *Ans.*

Annual Interest.

Art. 191.

(2.) Int. @ 8% on \$800 for 3 yr. = \$192.00
Int. @ 8% on \$800 for 1 yr. = \$64
Int. on annual int. 1 yr. = \$5.12
Int. on annual int. 3 yr. = \$15.36 . . 15.36
Total interest, \$207.36
Add principal, 800.00
Ans. \$1007.36

(3.) Int. on \$750 for 3 yr. @ 10% = \$225.00
Annual int. = \$75.00
Int. on annual int. = 7.50
Int. on an. int. 2 + 1, or 3, yr. = 22.50
\$247.50
Add principal, 750.00
Ans. \$997.50

(4.) Int. on $10000 for 4 yr. @ 5% = $2000.

Annual int. = $500

Int. on annual int. = 25

Int. on an. int. 3 + 2 + 1, or 6, yr. = 150.

Ans. $2150.

(5.)

yr.	mo.	da.	
1877	9	1	
1875	6	1	
2	3		$= 2\frac{1}{4}$ yr.

Int. on $500 for 1 yr. @ 6% = $30.00

Int. on $500 for $2\frac{1}{4}$ yr. @ 6% = 67.50 . . $67.50

Each semi-annual int. = 15.00

Interest on int. each half-yr. = .45

Interest on int. $3\frac{1}{2} + 2\frac{1}{2} + 1\frac{1}{2} + \frac{1}{2}$, or 8, half-yr. . . 3.60

Total interest, $71.10

Add principal, 500.00

Ans. $571.10

(6.)

yr.	mo.	da.	
1877	9	20	
1873	5	12	
4	4	8	$= 4\frac{16}{45}$ yr.

Int. on $1200 for 1 yr. @ 6% = $72.00

Int. on $1200 for $4\frac{16}{45}$ yr. @ 6% = 313.60 . . $313.60

Int. on annual int. 1 yr. = 4.32

Int. on an. int. $3\frac{16}{45} + 2\frac{16}{45} + 1\frac{16}{45} + \frac{16}{45}$, or $7\frac{19}{45}$, yr. = 32.06

Total interest, $345.66

Add principal, 1200.00

Ans. $1545.66

(7.)

yr.	mo.	da.	
1877	5	1	
1872	10	10	
4	6	21	$= 4\frac{67}{120}$ yr.

Int. on \$1500 for 1 yr. @ 5% = \$75.00
Int. on \$1500 for $4\frac{67}{120}$ yr. @ 5% = 341.88 . . \$341.88
Int. on an. int. 1 yr. = 3.75
Int. on an. int. $3\frac{67}{120} + 2\frac{67}{120} + 1\frac{67}{120} + \frac{67}{120}$, or
$8\frac{7}{30}$, yr. = 30.87
Total interest, \$372.75
Add principal, 1500.00
Ans. \$1872.75

(8.) Simple int. 1 yr. on \$1000 @ 6% = \$60: 5 yr. = \$300: int. on int. 1 yr. = \$3.60: for 4 + 3 + 2 + 1, or 10, yr. = \$36.00: annual int. = \$336; simple int. = \$300; difference = \$36, *Ans.*

(9.) \$500 × 6 = \$3000
Int. on \$3000 for 1 yr. @ 6% = \$180.00
Int. on \$3000 for 3 yr. @ 6% = 540.00 . . \$540.00
Int. on int. ½ yr. = 2.70
Int. on int. 5 + 4 + 3 + 2 + 1, or 15, half-yr. = 40.50
Total interest, \$580.50
Add principal, 3000.00
Ans. \$3580.50

(10.) Int. on \$20000 for 5 yr. @ 4% = \$4000.00
Int. on \$20000 for 1 yr. = \$800.00
Int. on \$20000 for ¼ yr. = 200.00
Int. on ¼ an. int. @ 1½% (¼ of 6%) = 3.00
Int. on ¼ an. int. 19+18+17+16+15+14+
13+12+11+10+9+8+7+6
+5+4+3+2+1, or 190, qrs. = 570.00
Total interest, \$4570.00
Add to this the premium on gold, 5%, = $\frac{1}{20}$, 228.50
Ans. \$4798.50

Art. 192.

yr.	mo.	da.	
1876	3	1	.. $44
1875	7	1	
	8 mo.		
1876	10	1	.. $10
1876	3	1	
	7 mo.		
1877	1	1	.. $26
1876	10	1	$36 ..
	3 mo.		
1877	12	1	.. $15
1877	1	1	
	11 mo.		
1878	3	16	
1877	12	1	
	3	15 = $3\frac{1}{2}$ mo.	

(2)

$350
 14 = int. 8 mo.
$364
 44
$320
 16 = int. 7 + 3 = 10 mo.
$336
 36
$300
 21.75 = int. 11 + $3\frac{1}{2}$ = $14\frac{1}{2}$ mo.
$321.75
 15.00
$306.75, *Ans.*

(3.) Amt. of $200, 1 yr. @ 6% = $212: $212 − $70 = $142: amt. of $142, 1 yr. @ 6% = $150.52, *Ans.*

(4)

1874	1	1	.. $109
1873	7	1	
	6 mo.		
1874	7	1	
1874	1	1	
	6 mo.		
1875	1	1	
1874	7	1	
	6 mo.		

6% per yr. = 3% per $\frac{1}{2}$ yr

$300	$6.00
.03	200
9.00	206
300	100
309	106
109	.03
200	3.18
.03	106
6.00	$109.18, *Ans*

(5)

1871	9	10	. . $32
1870	5	10	
1 yr. 4 mo.			
1872	9	10	. . $6.80
1871	9	10	
1 yr.			
1872	11	10	
1872	9	10	
	2 mo.		

$150
12 = int. for 1 yr. 4 mo.
162
32
130
9.10 = int. 1 yr. 2 mo.
139.10
6.80
$132.30, *Ans.*

(6)

1872	6	5	. . $20
1871	3	5	
1 yr. 3 mo.			
1872	12	5	. . $50.50
1872	6	5	$70.50
	6 mo.		
1874	6	5	
1872	12	5	
1 yr. 6 mo.			

$200
35 = int. 1 yr. 9 mo.
235
70.50
164.60
24.68 = int. 1 yr. 6 mo.
$189.18, *Ans.*

(7)

1875	6	1	. . $6
1875	1	1	
	5 mo.		
1876	1	1	. . $21.50
1875	6	1	$27.50
	7 mo.		
1876	7	1	
1876	1	1	
	6 mo.		

$250
17.50 = int. 12 mo.
267.50
27.50
240
8.40 = int. 6 mo.
$248.40, *Ans.*

(8)

1875	2	1 . . $25.40	
1874	8	1	
	6 mo.		
1875	8	1 . . $4.30	
1875	2	1	
	6 mo.		
1876	1	1 . . $30	
1875	8	1	$34.30
	5 mo.		
1876	7	1	
1876	1	1	
	6 mo.		

$180
5.40 = int. 6 mo.
185.40
25.40
160
8.80 = int. 11 mo.
168.80
34.30
134.50
4.035 = int. 6 mo.
$138.54, *Ans.*

(9)

1875	9	1 . . $10	
1875	3	1	
	6 mo.		
1876	1	1 . . $30	
1875	9	1	$40.
	4 mo.		
1876	7	1 . . $11	
1876	1	1	
	6 mo.		
1876	9	1 . . $80	
1876	7	1	$91.
	2 mo.		

1877	3	1
1876	9	1
	6 mo.	

$400
20 = int. 10 mo. (6+4.)
420
40
380
15.10 = int. 8 mo. (6+2.)
395.20
91
304.20
9.126 = int. 6 mo.
$313.33, *Ans.*

(10)

1877	1	1 . . $20		1877	12	25 . . $10	
1876	4	16		1877	7	16	
	8 mo.	15 da.			5 mo.	9 da.	
1877	4	1 . . $14		1878	7	4 . . $18	
1877	1	1		1877	12	25	$28
	3 mo.				6 mo.	9 da.	
1877	7	16 . . $31		1879	6	1	
1877	4	1	$65	1878	7	4	
	3 mo.	15 da.			10 mo.	27 da.	

$450 + $45 (int. 8 mo. 15 da. + 3 mo. + 3 mo. 15 da.) = $495: $495 — $65 = $430: $430 + $64.50 (int. 5 mo. 9 da. + 6 mo. 9 da. + 10 mo. 27 da.) = $494.50: $494.50 — $28 = $466.50, *Ans.*

(11)

1870	5	1 . . $18		1871	12	23 . . $324
1870	1	1		1871	7	13
	4 mo.				5 mo.	10 da.
1870	9	4 . . $20		1873	10	1
1870	5	1		1871	12	23
	4 mo.	3 da.		1 yr.	9 mo.	8 da.

1870	12	16 . . $15		$1000
1870	9	4		92 = int. 18 mo. 12 da.
	3 mo.	12 da.		1092
1871	4	10 . . $21		192
1870	12	16		900
	3 mo.	24 da.		24 = int. 5 mo. 10 da.
1871	7	13 . . $118		924
1871	4	10	$192	324
	3 mo.	3 da.		600
				63.80 = int. 1 yr. 9 mo. 8 da.
				$663.80, *Ans.*

Art. 193.

(1.) Int. \$320, 1 yr. @ 6% = \$19.20
Amount = \$339.20
Amt. of \$50, 8 mo. @ 6% = \$52.00
Amt. of \$100, $1\frac{1}{2}$ mo. @ 6% = 100.75 152.75
Balance due, \$186.45, *Ans.*

(2.) Time from March 1, 1877, to Jan. 1, 1878, = 10 mo.
Amt. of \$540, 10 mo. @ 8% = \$576.00
Amt. of \$90, 8 mo. = \$94.80
Amt. of \$100, 6 mo. = 104.00
Amt. of \$150, 5 mo. = 155.00
Amt. of \$180, 2 mo. 20 da. = 183.20 537.00
Balance due, \$39.00, *Ans.*

DISCOUNT.

Case I.

Art. 196.

1st. *When the note does not bear interest.*

(2.) Days in June, 10
Days in July, 31
Days in Aug., 19
60

Int. of \$1 for 63 da. @ 6% = \$0.0105.
\$100 × .0105 = \$1.05
\$100 — \$1.05 = \$98.95

To 19th Aug. add 3 da. grace.

Ans. Aug. 19/22. \$1.05, \$98.95

(3.) Remaining days in Oct., 19: 30—19 = Nov. 11/14: int. on \$1 33, da. @ 8% = \$0.0073+: \$120 × .0073+ = \$0.88: \$120 — \$0.88 = \$119.12

Ans. Nov. 11/14, \$0.88, \$119.12

(4.) Int. of \$1, 4 mo. 3 da. @ 6% = \$0.0205 : \$140 × .0205 = \$2.87 : \$140 — \$2.87 = \$137.13.

Ans. May $^{15}/_{18}$, \$2.87, \$137.13

(5.) Int. of \$180, 1 yr. @ 4% = \$7.20 : int. of \$180, 6 mo. @ 4% = \$3.60 : int. of \$180, 3 da. @ 4% = \$0.06 : \$3.60 + \$0.06 = \$3.66 : \$180 — \$3.66 = \$176.34

Ans. Oct. $^{10}/_{13}$, \$3.66, \$176.34

(6.) Int. of \$250, 1 yr. @ 8% = \$20.00 : int. of \$250, 5 mo. 3 da. = \$8.50 : \$250 — \$8.50 = \$241.50

Ans. May $^{1}/_{4}$, \$8.50, \$241.50

(7.) Days remaining in Aug., 27, + 3 = Sept. $^{3}/_{6}$: 6% on \$1 for 33 da. = \$0.0055 : \$375 × .0055 = \$2.06 : \$375 — \$2.06 = \$372.94 *Ans.* Sept. $^{3}/_{6}$, \$2.06, \$372.94

(8.) Int. on \$600 for 2 mo. 3 da. = \$9.45 : \$600 — \$9.45 = \$590.55 *Ans.* Apr. $^{12}/_{15}$, \$9.45, \$590.55

(9.) Remaining days in Feb., 8, March, 31, April, 30 = 69 da.: 90 — 69 = May $^{21}/_{24}$. Int. on \$1200, 1 mo. @ 10% = \$10 : for $3\frac{1}{10}$ mo. = \$31 : \$1200 — \$31 = \$1169.

Ans. May $^{21}/_{24}$, \$31, \$1169.

(10.) Int. on \$1, 93 da. @ 6% = \$0.0155 : \$1780 × .0155 = \$27.59 : days remaining in Jan., 20, + 29 (Feb., leap yr.) + 31 (Mar.) = 80 : 90 — 80 = Apr. $^{10}/_{13}$. \$1780 — \$27.59 = \$1752.41 *Ans.* Apr. $^{10}/_{13}$, \$27.59, \$1752.41

(11.) Due Sept. $^{15}/_{18}$, 1877: number of days from May 21 to Sept. 18 = May, 10, June, 30, July, 31, Aug., 31, Sept., 18 = 120: int. on $600 for 120 da. (4 mo.) at 10% = $20: $600 — $20 = $580.

Ans. Sept. $^{15}/_{18}$, 1877, 120 da., $20, $580.

(12.) In May, 23 da., June, 30, July, 31 = 84: 90 — 84 = $^{6}/_{9}$ Aug.: June 8 to Aug. 9 = 48 da. = $1\frac{3}{5}$ mo.: int. on $1000, $1\frac{3}{5}$ mo. @ 6% = $8: $1000 — $8 = $992. *Ans.* Aug. $^{6}/_{9}$, 48 da., $8, $992.

(13.) 6 mo. after July 10, 1877, = Jan. $^{10}/_{13}$, 1878: days in Oct., 7*, Nov., 30, Dec., 31, Jan., 13 = 81: 81 da. = 2 mo. 21 da.: int. on $1500 for this time @ 6% = $20.25: $1500 — $20.25 = $1479.75

Ans. Jan. $^{10}/_{13}$, 1878, 81 da., $20.25, $1479.75

2D. *When the note bears interest.*

(2.) 6 mo. from May 20, 1875, = Nov. $^{20}/_{23}$: amt. of $150 @ 6% int. 6, mo. 3 da. = $154.58: Sept. 9 to Nov. 23 = 75 da. or $2\frac{1}{2}$ mo.: discount on $154.58, $2\frac{1}{2}$ mo. @ 8% = $2.58: $154.58 — $2.58 = $152.

Ans. Nov. $^{20}/_{23}$, 1875, 75 da., $2.58, $152.

---

* See Rem. 3, page 249, Ray's New Practical Arithmetic.

(3.) 1 yr. from Aug. 5, 1876, = Aug. $^5/_8$, 1877: amt. of $300, 1 yr. 3 da. @ 8% int. = $324.20: Apr. 16 to Aug. 8 = 114 da. or $3\frac{4}{5}$ mo.: discount on $324.20, $3\frac{4}{5}$ mo. at 6% = $6.16: $324.20 — $6.16 = $318.04 = proceeds.

Ans. Aug. $^5/_8$, 1877, 114 da., $6.16, $318.04

(4.)

1878	1	4
1877	3	4
	10 mo.	

$450, 10 mo. @ 6% amounts to $472.50: Aug. 13, 1877, to Jan. 4, 1878, = 144 da. or $4\frac{4}{5}$ mo.: discount on $472.50, $4\frac{4}{5}$ mo. @ 10% = $18.90: $472.50 — $18.90 = $453.60

Ans. Jan. $^1/_4$, 1878, 144 da., $18.90, $453.60

(5.)

1878	9	4
1876	5	16
2	3	18

$650, 2 yr. 3 mo. 18 da. @ 9% = $784.55: Apr. 25, 1878, to Sept. 4, 1878, = 132 da. or $4\frac{2}{5}$ mo.: discount @ 6% on $784.55 for $4\frac{2}{5}$ mo. = $17.26: $784.55 — $17.26 = $767.29 = proceeds.

Ans. Sept. $^1/_4$, 1878, 132 da., $17.26, $767.29

(6.) Amt. of $840, 6 mo. 3 da. @ 10% = $882.70: Dec. 20, 1875, to Mar. 4, 1876 = 75 da. or $2\frac{1}{2}$ mo.: discount on $882.70, $2\frac{1}{2}$ mo. @ 8% = $14.71: $882.70 — $14.71 = $867.99 = proceeds.

Ans. Mar. $^1/_4$, 1876, 75 da., $14.71, $867.99

(7.) 1876 5 4
1875 7 19
9 mo. 15 da.

Amt. of \$1400, $9\frac{1}{2}$ mo. @ 6% = \$1466.50: Jan. 17, 1876, to May 4, 1876, = 108 da. or $3\frac{3}{5}$ mo.: discount on \$1466.50, $3\frac{3}{5}$ mo. @ 10% = \$44.00: \$1466.50 — \$44.00 = \$1422.50 = proceeds.

Ans. May $^{1}/_{4}$, 1876, 108 da., \$44, \$1422.50

(8.) 1878 1 4
1876 10 16
1 yr. 2 mo. 18 da.

Amt. of \$2400, 1 yr. 2 mo. 18 da. @ 8% = \$2633.60: July 26, 1877, to Jan. 4, 1878 = 162 da. or $5\frac{2}{5}$ mo.: discount on \$2633.60, $5\frac{2}{5}$ mo. @ 10% = \$118.51: \$2633.60 — \$118.51 = \$2515.09 = proceeds.

Ans. Jan. $^{1}/_{4}$, 1878, 162 da., \$118.51, \$2515.09

(9.) Amt. of \$3500 @ 6%, 1 yr. 3 da. = \$3711.75: May 15 to Oct. 18, 1878, = 156 da. or $5\frac{1}{5}$ mo.: discount on \$3711.75, $5\frac{1}{5}$ mo. @ 9% = \$144.76: \$3711.75 — \$144.76 = \$3566.99 = proceeds.

Ans. Oct. $^{15}/_{18}$, 1878, 156 da., \$144.76, \$3566.99

(10.) Amt. of \$6000, 1 yr. 3 da. @ 8% = \$6484.00: Nov. 21, 1875, to May 13, 1876, = 174 da. or $5\frac{4}{5}$ mo.: discount on \$6484, $5\frac{4}{5}$ mo. @ 10% = \$313.39: \$6484.00 — \$313.39 = \$6170.61 = proceeds.

Ans. May $^{10}/_{13}$, 1876, 174 da., \$313.39, \$6170.61

Art. 197.

(2.) Bank discount on \$1, 63 da. @ 6% = \$0.0105: \$1 — \$0.0105 = \$0.9895: \$197.90 ÷ .9895 = \$200, *Ans.*

(3.) Discount on \$1, 93 da. @ 6% = \$0.0155: \$1 — \$0.0155 = \$0.9845: \$393.80 ÷ .9845 = \$400, *Ans.*

(4.) Discount on \$1, 5 mo. 3 da. @ 8% = \$0.034: \$1 — \$0.034 = \$0.966: \$217.35 ÷ .966 = \$225, *Ans.*

(5.) Discount on \$1, 4 mo. 3 da. @ 6% = \$0.0205: \$1 — \$0.0205 = \$0.9795: \$352.62 ÷ .9795 = \$360, *Ans.*

(6.) Discount on \$1, 33 da. @ 6% = \$0.0055: \$1 — \$0.0055 = \$0.9945: \$400 ÷ .9945 = \$402.21+, *Ans.*

(7.) Discount on \$1, 2 mo. 3 da. @ 8% = \$0.014: \$1 — \$0.014 = \$0.986: \$500 ÷ .986 = \$507.10 (nearly), *Ans.*

(8.) Discount on \$1, 6 mo. 3 da. @ 10% = \$0.050833+: \$1 — \$0.050833 = \$0.949166: \$1500 ÷ .949166 = \$1580.33+, *Ans.*

(9.) Oct. 12, 1876, to Jan. 4, 1877, = 2 mo. 24 da., or $2\frac{4}{5}$ mo.: discount on \$1, $2\frac{4}{5}$ mo. @ 6% = \$0.014: \$1 — \$0.014 = \$0.986: \$1055.02 ÷ .986 = \$1070.

1877	1	4
1876	2	19
	10 mo.	15 da.

\$1 @ 8% for 10 mo. 15 da. = \$0.07: \$1 + \$0.07 = \$1.07: \$1070 ÷ 1.07 = \$1000, *Ans.*

Art. 199.

(3.) Amt. of \$1, 2 yr. @ 6% = \$1.12: \$224 ÷ 1.12 = \$200 = present worth: \$224 — \$200 = \$24 = discount.

(4.) Amt. of \$300 for 2 yr. @ 8% = \$348: amt. of \$1 for 2 yr. @ 6% = \$1.12: \$348 ÷ 1.12 = \$310.71 = present worth: \$348 — \$310.71 = \$37.29 = discount.

(5.) Amt. of \$1, 5 yr. 10 mo. @ 6% = \$1.35: \$675 ÷ 1.35 = \$500 = present worth: \$675 — \$500 = \$175 = discount.

(6.) Amt. of \$1, 5 mo. @ 10% = \$1.04166+: \$368.75 ÷ 1.04166 = \$354 = present worth: \$368.75 — \$354 = \$14.75 = discount.

(7.)

1878	1	1
1876	9	10
1	3	21

Amt. of \$800, 1 yr. 3 mo. 21 da. @ 6% = \$862.80: July 19, 1877, to Jan. 1, 1878 = 5 mo. 12 da.: amt. of \$1, 5 mo. 12 da. @ 10% = \$1.045: \$862.80 ÷ 1.045 = \$825.65 = present worth: \$862.80 — \$825.65 = \$37.15 = discount.

(8.) Amt. of \$1, 4 mo. @ 10% = \$1.03⅓: \$775 ÷ 1.03⅓ = \$750, *Ans.*

(9.) Amt. of \$1, 8 mo. @ 6% = \$1.04: \$260 ÷ 1.04 = \$250, *Ans.*

(10.) \$2480 — its 5% (\$124) = \$2356 = cash cost. Amt. of \$1, 4 mo. @ 10% = \$1.03⅓: \$2480 ÷ 1.03⅓ = \$2400 present worth. \$2400 — \$2356 = \$44, *Ans.*

(11.) ⅓ of \$956.34 = \$318.78
Amt. of \$1, 1 yr. @ 5% = \$1.05: \$318.78 ÷ 1.05 = \$303.60
Amt. of \$1, 2 yr. @ 5% = \$1.10: \$318.78 ÷ 1.10 = 289.80
Amt. of \$1, 3 yr. @ 5% = \$1.15: \$318.78 ÷ 1.15 = 277.20
Ans. \$870.60

(12.) \$535 × .07 = \$37.45 = bank discount: \$535 ÷ 1.07 = \$500: \$535 — \$500 = \$35: \$37.45 — \$35 = \$2.45, *Ans.*

(13.) \$750—(\$750 × .04) = \$720, cash cost. Amt. of \$1 for 3 mo. @ 8% = \$1.02: \$750 ÷ \$1.02 = \$735.29, present worth: \$735.29 — \$720 = \$15.29, *Ans.*

(14.) ⅓ of \$10296 = \$3432. Amt. of \$1, 1 yr. @ 10% = \$1.10: amt. of \$1, 2 yr. @ 10% = \$1.20: amt. of \$1, 3 yr. @ 10% = \$1.30.

\$3432 ÷ 1.10 = \$3120
\$3432 ÷ 1.20 = 2860
\$3432 ÷ 1.30 = 2640 \$8620
\$8620 — \$8000 = \$620, *Ans.*

(15.) July 4, 1876, to May 1, 1878, = 1 yr. 9 mo. 27 da. Amt. of \$2000, 1 yr. 9 mo. 27 da. @ 8% = \$2292: Oct. 25, 1877, to May 1, 1878, = 6 mo. 6 da.: Amt. of \$1, 6 mo. 6 da. @ 6% = \$1.031: \$2292 ÷ 1.031 = \$2223.08 = present worth: \$2292 — \$2223.08 = \$68.92 = discount.

EXCHANGE.

Art. 201.

(1.) 1% of \$1400 = \$14: $\frac{1}{2}$% = \$$\frac{14}{2}$ = \$7: \$1400 + \$7 = \$1407, *Ans.*

(2.) $\frac{1}{2}$% of \$2580 = \$12.90: \$2580 — \$12.90 = \$2567.10, *Ans.*

(3.) \$375.87 = 100% + $\frac{1}{8}$% of the face: \$375.87 ÷ 100$\frac{1}{8}$ = \$375.40, *Ans.*

(4.) $\frac{1}{4}$% of \$2785 = \$6.96: \$2785 — \$6.96 = \$2778.04, *Ans.*

(5.) 100% — 1$\frac{1}{4}$ = 98$\frac{3}{4}$% = .9875: \$1852.55 ÷ .9875 = \$1876, *Ans.*

(6.) Int. of \$5680. for 63 da. @ 6% = \$59.64; $\frac{1}{2}$ % prem. on \$5680. = \$28.40; \$59.64 — \$28.40 = \$31.24; \$5680. — \$31.24 = \$5648.76, *Ans.*

(7.) Int. of \$1575. for 33 da. @ 6% = \$8.66; $\frac{3}{4}$ % prem. on \$1575. = \$11.81; \$11.81 — \$8.66 = \$3.15; \$1575. + \$3.15 = \$1578.15, *Ans.*

(8.) Int. of \$2625. for 63 da. @ 6% = \$27.56; 1$\frac{1}{2}$% prem. on \$2625. = \$39.37; \$39.37 — \$27.56 = \$11.81; \$2625. + 11.81 = \$2636.81, *Ans.*

Art. 202.

(3.) 8s. $= \frac{4}{10}$£: £890.4 × 4.86 ($) = $4327.34, *Ans.*

(4.) $2130.12 ÷ 4.88 = 436, with 244 rem.: 244 × 20s., ÷ 4.88 = 10. *Ans.* £436 10s.

(5.) 5 fr. 15 centimes = 5\frac{3}{20}$ fr.: 1290 ÷ 5$\frac{3}{20}$ = $250.49, *Ans.*

(6.) $1657.60 × 5$\frac{16}{100}$ = 8553 fr. 22, *Ans.*

(7.) $12680 ÷ 4 = 3170, × .97 = $3074.90, *Ans.*

(8.) If 4 marks = $0.98, 1 m. = $0.245: $1470 ÷ .245 = 6000 m., *Ans.*

INSURANCE.

Art. 204.

(2.) $\frac{3}{4}$ of $5000 = $3750: $\frac{1}{2}$% of $3750 = $18.75, add $1.50 = $20.25, *Ans.*

(3.) $\frac{2}{3}$ of $12600 = $8400, @ $\frac{3}{4}$% = $63.00
$\frac{1}{2}$ of $14400 = $7200, @ 2% = 144.00
2 policies @ $1.25 = 2.50
Ans. $209.50

(4.) $\frac{4}{7}$ of $21000 = $12000, @ 1$\frac{1}{2}$% = $180.00
$7200 @ $\frac{3}{4}$% = 54.00
2 policies @ $1.25 = 2.50
Ans. $236.50

(5.) $\frac{3}{4}$ of $5600 = $4200, × .01$\frac{1}{2}$ = $63, × 20 (yr.) = $1260: $4200 — $1260 = $2940, *Ans.*

(6.) $3600 + $1600 + $800 = $6000: $\frac{7}{8}$% of $6000 = $52.50, + $1.25 = $53.75, *Ans.*

(7.) \$151.25 — \$1.25 = \$150: \$150 = $1\frac{1}{2}\%$ of $\frac{2}{3}$ value: \$100 = 1% of $\frac{2}{3}$ value: \$10000 = 100% of $\frac{2}{3}$ value: $\frac{3}{3}$, or the whole value, = \$15000, *Ans.*

(8.) $\frac{4}{5}$ of \$4500 = \$3600: \$32.75 — \$1.25 = \$31.50: \$31.50 ÷ 3600 = $.0087\frac{1}{2} = \frac{7}{8}\%$, *Ans.*

(9.) \$1000 + \$1500 = \$2500: \$3.50 ÷ 2500 = .0014 = $\frac{7}{50}\%$, *Ans.*

Art. 205.

(2.) \$105.53 × 10 = \$1055.30 = amount paid yearly: \$1055.30 × 10 = \$10553, *Ans.*

(3.) \$47.18 × 8 × 20 = \$7548.80: \$60.45 × 8 × 20 = \$9672.00: \$9672 — \$7548.80 = \$2123.20, *Ans.*

(4.) \$36.46 × 12 × 5 = \$2187.60: \$12000 — \$2187.60 = \$9812.40, *Ans.*

(5.) 75 yr. — 21 yr. = 54 yr.: \$19.89 × 5 × 54 = \$5370.30, *Ans.*

(6.) \$104.58 × 10 (yr.) = \$1045.80 There will be int. @ 6% on \$104.58, 10 + 9 + 8 + 7 + 6 + 5 + 4 + 3 + 2 + 1, or 55, yr. = \$345.11: \$1045.80 + \$345.11 = \$1390.91, *Ans.*

(7.) \$29.15 × 6 = \$174.90, × 15 = \$2623.50: int. @ 6% on \$174.90 for 15 + 14 + 13 + 12 + 11 + 10 + 9 + 8 + 7 + 6 + 5 + 4 + 3 + 2 + 1, or 120, yr. = \$1259.28, + \$2623.50 = \$3882.78, *Ans.*

TAXES.

Art. 208.

(2.) \$2500 — \$28 = \$2472: 2472 ÷ 618000 = .004 *Ans.* 4 mills on \$1, or $\frac{2}{5}\%$.

(3.) 18409.44 ÷ 2876475 = .0064 = 6.4 mills = *Ans.*

(4.) 656491.61 ÷ 421285359 = .00156 = 1.56 mills = *Ans.*

Art. 209.

[I.] $1.25 × 57 = $71.25 : $1373.64 — $71.25 = $1302.39 : 1302.39 ÷ 748500 = .00174 = rate 1.74 mills on $1.

(2.) $2576 × .00174 = $4.48, + $1.25 (poll-tax) = $5.73, *Ans.*

(3.) $9265 × .00174 = $16.12, + $3.75 (3 poll-taxes) = $19.87, *Ans.*

(4.) $4759 × .00174 = $8.28, + $1.25 = $9.53, *Ans.*

(5.) $8367 × .00174 = $14.56, *Ans.*

[II.] 64375 ÷ 16869758 = .003816. Rate 3.816 mills on $1.

TAX TABLE.—Rate, 3.816 mills on $1.

PROP.	TAX.	PROP.	TAX.	PROP.	TAX.	PROP.	TAX.
$1	$0.004	$10	$0.038	$100	$0.382	$1000	$3.816
2	.008	20	.076	200	.763	2000	7.632
3	.011	30	.114	300	1.145	3000	11.448
4	.015	40	.153	400	1.526	4000	15.264
5	.019	50	.191	500	1.908	5000	19.080
6	.023	60	.229	600	2.290	6000	22.896
7	.027	70	.267	700	2.671	7000	26.712
8	.030	80	.305	800	3.053	8000	30.528
9	.034	90	.343	900	3.434	9000	34.344

(1.) $56875 × .003816 = $217.04, *Ans.*

(2.) $27543 × .003816 = $105.10, *Ans.*

(3.) $83612 × .003816 = $319.06, *Ans.*

(4.) $72968 × .003816 = $278.45, *Ans.*

(5.) $69547 × .003816 = $265.39, *Ans.*

Art. 211.

(1.) 36 sq. mi. contain 23040 A.: 23040 A. @ $1.25 per acre = $28800, *Ans.*

(2.) The charge will be the same as for 3 half-ounces. 3 times 3 ct. = 9 ct., *Ans.*

(3.) 1 lb. 5 oz. = 21 oz.: postage same as for 22 oz.: 22 ÷ 2 = 11: 11 times 1 ct. = 11 ct., *Ans.*

(4.) 70 ct. times 40 = $28, *Ans.*

(5.) $5 = 500 ct.: $\frac{500}{1000}$ ct. = $\frac{1}{2}$ ct., *Ans.*

(6.)	30000 bl. @ $1	=	$30000
	250 ret. dlrs. @ $20 ea.	=	5000
	12 wholesale dlrs. @ $50	=	600
			$35600, *Ans.*

Art. 212.

(1.) $12\frac{1}{2}\%$ = $\frac{1}{8}$: 1760 lb. — its $\frac{1}{8}$ = 1540 lb.: 1540 times 0.01\frac{3}{4}$ = $26.95, *Ans.*

(2.) 40 bales of 400 lb. each = 16000 lb.: 5% tare = 800 lb.: 16000 — 800 = 15200: 15200 lb. @ 45 ct. = $6840: 10% ad. val. = $684: 15200 lb. @ 9 ct. duty = $1368: $684 + $1368 = $2052, *Ans.*

(3.) 365.15 fr. + 57.15 fr. = 422.30 fr., to which add 5% com. (21.1150 fr.) = 443.4150 fr.: 443.4150 × 19$\frac{3}{10}$ (ct.) = $85.58: 40% of $86 = $34.40, *Ans.*

(4.) 1317.04 mk. + 34.36 mk. = 1351.40 mk.: add 6% com. (81.084 mk.) = 1432.484 mk.: 1432.484 × 23.8 (ct.) = $340.93: 25% of $341 = $85.25, *Ans.*

(5.) 50 ct. per lb. duty on 1500 lb. $= \$750$: £8 4s. 6d. $=$ £$8\frac{9}{40}$ or £8.225: £500 $+$ £8.225 $=$ £508.225: add $2\frac{1}{2}\%$ com. (£12.705$+$) $=$ £520.93: £520.93 $\times$ 4.8665 (\$) $=$ \$2535.11: 35% of \$2535 $=$ \$887.25: \$887.25 $+$ \$750 $=$ \$1637.25, *Ans.*

RATIO.

Art. 214.

(20.) $\frac{7}{2} \times \frac{4}{9} = \frac{14}{9} = 1\frac{5}{9}$, *Ans.*

(21.) $\frac{35}{6} \times \frac{3}{7} = \frac{5}{2} = 2\frac{1}{2}$, *Ans.*

(22.) $\frac{69}{10} \times \frac{5}{23} = \frac{3}{2} = 1\frac{1}{2}$, *Ans.*

(28.) 5 yd. 1 ft. $= 192$ in.: 5 ft. 4 in. $= 64$ in.: $\frac{192}{64} = 3$, *Ans.*

Art. 215.

(8.) 4 lb. 8 oz. $= 72$ oz.: $\frac{7}{8}$ of 72 oz. $= 63$ oz.: 63 oz. $= 3$ lb. 15 oz., *Ans.*

(9.) $\$4.00 \times 2.6 = \$10.40$, *Ans.*

Art. 216.

(3.) $42 \times \frac{10}{7} = 60$, *Ans.*

(4.) $23\frac{3}{8} = \frac{187}{8}$: $\frac{187}{8} \times \frac{4}{11} = \frac{17}{2} = 8\frac{1}{2}$, *Ans.*

(5.) $7\frac{5}{9} = \frac{68}{9}$: $\$27.20 \times \frac{9}{68} = \$0.40 \times 9 = \$3.60$, *Ans.*

Art. 217.

(2.) $\left.\begin{array}{r} 5 \times 6 = 30 \\ 10 \times 9 = 90 \end{array}\right\} 90 \div 30 = 3$, *Ans.*

(3.)
$$\begin{array}{c|c} & 2 \\ \not{6}\frac{1}{\not{4}} & \not{1}\not{2}\frac{1}{\not{2}} \\ \not{8}\frac{1}{\not{3}} & \not{3}\not{3}\frac{1}{\not{3}} \\ & 4 \end{array} \quad 4 \times 2 = 8, \textit{ Ans.}$$

(4.) $\frac{1}{2} \times \frac{5}{4} = \frac{5}{8}$: $\frac{2}{3} \times \frac{3}{4} = \frac{6}{12} = \frac{1}{2}$: $\frac{\frac{1}{2}}{\frac{5}{8}} = \frac{8}{10} = \frac{4}{5}$, *Ans.*

(5.) $2 \times 24 = 48$: $8 \times 12 = 96$: $96 \div 48 = 2$, *Ans.*

(6.)
$$\begin{array}{r|rr} \$\not{2}.\not{2}\not{5} & \$\not{6}.\not{7}\not{5} & 3 \\ \not{3} & \not{6} & 2 \end{array} \qquad 3 \times 2 = 6, \textit{ Ans.}$$

(7.)
$$\begin{array}{r|r} 2 & \not{5} \\ \not{3} & 7 \\ \not{5} & \not{9} \\ & 3 \end{array} \qquad \frac{7 \times 3}{2} = \frac{21}{2} = 10\tfrac{1}{2}, \textit{ Ans.}$$

Art. 219.

(2.) Divide by 5.

(3.) Divide by 10.

(4.) Divide by 17.

(5.) Divide by 19.

(6.) Divide by 25.

(7.) Divide by 31.

Art. 220.

(2.)
$$\begin{array}{rcr} 3\frac{3}{4} & : & 4\frac{2}{5} \\ & & 20 \\ \hline 75 & : & 88, \textit{ Ans.} \end{array}$$

(3.)
$$\begin{array}{rcr} 7\frac{1}{2} & : & 10\frac{2}{3} \\ & & 6 \\ \hline 45 & : & 64, \textit{ Ans.} \end{array}$$

(4.) $\frac{5}{6} = \frac{15}{18}$: $\frac{7}{9} = \frac{14}{18}$. 15 : 14, *Ans.*

(5.) $\frac{63}{10} = \frac{189}{30}$: $9\frac{7}{15} = \frac{142}{15} = \frac{284}{30}$. 189 : 284, *Ans.*

PROPORTION.

Art. 223.

(3.)
$$\frac{\overset{4}{\not{8}} \times 6}{\not{2}} = 24, \textit{ Ans.}$$

(4.)
$$\frac{7 \times \overset{2}{\not{1}\not{0}}}{\not{5}} = 14, \textit{ Ans.}$$

(5.)
$$\frac{\not{8} \times \overset{3}{\not{6}}}{\underset{2}{\not{1}\not{6}}} = 3, \textit{ Ans.}$$

(6.)
$$\frac{5 \times \overset{2}{\not{1}\not{2}}}{\not{6}} = 10, \textit{ Ans.}$$

(7.) $\dfrac{3 \times \overset{2}{\not{14}}}{\not{7}} = 6$, *Ans.*

(8.) $\dfrac{\overset{2}{\not{14}} \times 9}{\not{7}} = 18$. *Ans.*

(9.) $\dfrac{2 \times \overset{2}{\not{8}} \times \overset{5}{\not{45}}}{\not{4} \times \not{9}} = 20$, *Ans.*

(10.) $\dfrac{\overset{\not{2}}{\not{8}} \times \overset{\not{2}}{\not{10}} \times 3 \times 7}{\not{5} \times \not{4} \times \underset{\not{2}}{\not{4}}} = 21$, *Ans.*

(11.) $\dfrac{\overset{5}{\not{10}} \times \overset{\not{2}}{\not{14}} \times \overset{\not{3}}{\not{33}} \times \overset{3}{\not{39}}}{\underset{\not{3}}{\not{21}} \times \underset{\not{2}}{\not{22}} \times \underset{\not{2}}{\not{26}}} = 15$, *Ans.*

(12.) $\dfrac{3}{\not{4}} \times \dfrac{\not{4}}{5} = \frac{3}{5}$: $\frac{3}{5} \times \frac{3}{2} = \frac{9}{10}$, *Ans.*

(13.) $\dfrac{3}{\not{5}} \times \dfrac{\not{5}}{4} = \frac{3}{4}$: $\dfrac{\not{3}}{\underset{2}{\not{4}}} \times \dfrac{\overset{5}{\not{10}}}{\underset{3}{\not{9}}} = \frac{5}{6}$, *Ans.*

(14.) $\dfrac{\overset{7}{\not{14}}}{\not{3}} \times \dfrac{\overset{5}{\not{15}}}{\not{2}} = 35$: $\dfrac{\overset{5}{\not{35}}}{1} \times \dfrac{2}{\underset{3}{\not{21}}} = \frac{10}{3} = 3\frac{1}{3}$, *Ans.*

(15.) $\frac{6 \times 6}{4} = 9$, *Ans.*

Art. 224.

(3.) 6 : 12 : : 3 : ? $\qquad \dfrac{\overset{2}{\not{12}} \times 3}{\not{6}} = 6$, *Ans.*

(4.) 3 : 6 : : \$8 : ? $\qquad \dfrac{\overset{2}{\not{6}} \times 8}{\not{3}} = \$16$, *Ans.*

(5.) $5 : 3 :: \$30 : ?$ $\qquad \frac{3 \times \overset{6}{\cancel{30}}}{\cancel{5}} = \$18$, *Ans.*

(6.) 3 lb. 12 oz. = 60 oz.: 11 lb. 4 oz. = 180 oz.

$60 : 180 :: \$3.50 : ?$ $\qquad \frac{\overset{3}{\cancel{180}} \times 3.50}{\cancel{60}} = \$10.50$, *Ans.*

(7.) 2 lb. 8 oz. = 40 oz. $\$2 : \$5 :: 40 \text{ oz.} : ?$

$\frac{5 \times \overset{20}{\cancel{40}}}{\cancel{2}} = 100$ oz. = 6 lb. 4 oz., *Ans.*

(8.) $4 : 10 :: \$14 : ?$ $\qquad \frac{\overset{5}{\cancel{10}} \times \overset{7}{\cancel{14}}}{\underset{2}{\cancel{4}}} = \$35$, *Ans.*

(9.) $3 : 11 :: 69 \text{ ct.} : ?$ $\qquad \frac{11 \times \overset{23}{\cancel{69}}}{\cancel{3}} = \$2.53$, *Ans.*

(10.) $4 : 9 :: \$7 : ?$ $\qquad \frac{9 \times 7}{4} = \frac{63}{4} = \$15.75$, *Ans.*

(11.) $8 : 12 :: \$32 : ?$ $\qquad \frac{12 \times \overset{4}{\cancel{32}}}{\cancel{8}} = \$48$, *Ans.*

(12.) $12 : 8 :: \$48 : ?$ $\qquad \frac{\overset{4}{\cancel{48}} \times 8}{\cancel{12}} = \$32$, *Ans.*

(13.) $\$32 : \$48 :: 8 : ?$ $\qquad \frac{\overset{12}{\cancel{48}} \times \cancel{8}}{\underset{4}{\cancel{32}}} = 12$ yd., *Ans.*

(14.) \$48 : \$32 : : 12 : ?

$$\frac{\overset{8}{\cancel{32}} \times \cancel{12}}{\underset{\cancel{4}}{\cancel{48}}} = 8 \text{ yd., } \textit{Ans.}$$

(15.) 19 : 4 : : \$152 : ?

$$\frac{4 \times \overset{8}{\cancel{152}}}{\cancel{19}} = \$32, \textit{ Ans.}$$

(16.) 12 : 8 : : 24 : ?

$$\frac{8 \times \overset{2}{\cancel{24}}}{\cancel{12}} = 16 \text{ da., } \textit{Ans.}$$

(17.) 2 : 8 : : 60 : ?

$$\frac{\overset{4}{\cancel{8}} \times 60}{\cancel{2}} = 240 \text{ men, } \textit{Ans.}$$

(18.) 6 lb. = 96 oz. 15 : 96 : : 25 ct. : ?

$$\frac{\overset{32}{\cancel{96}} \times \overset{5}{\cancel{25}}}{\underset{\cancel{3}}{\cancel{15}}} = \$1.60, \textit{ Ans.}$$

(19.) 6 : 26 : : \$2.70 : ?

$$\frac{26 \times \overset{.45}{\cancel{2.70}}}{\cancel{6}} = \$11.70, \textit{ Ans.}$$

(20.) 585 lb. : 3525 lb. : : \$42.12 : ?

$$\frac{\overset{705}{\cancel{3525}} \times \overset{.36}{\cancel{42.12}}}{\underset{\cancel{117}}{\cancel{585}}} = \$253.80, \textit{ Ans.}$$

(21.) $\frac{3}{2}$: $\frac{9}{8}$: : \$2.50 : ?

$$\frac{\overset{3}{\cancel{9}} \times 2.50}{\underset{4}{\cancel{8}}} \times \frac{\cancel{2}}{\cancel{3}} = \$1.87\tfrac{1}{2}, \textit{ Ans.}$$

(22.) 90 : 450 : : 6 : ?

$$\frac{\overset{30}{\cancel{450}} \times \cancel{6}}{\underset{\cancel{6}}{\cancel{90}}} = 30 \text{ da., } \textit{Ans.}$$

(23.) 5 : 15 : : 6 : ?

$$\frac{\overset{3}{\cancel{15}} \times 6}{\cancel{5}} = 18 \text{ men, } \textit{Ans.}$$

(24.) 30 : 140 : : 15 : ?

$$\frac{\overset{70}{\cancel{140}} \times \cancel{15}}{\underset{\cancel{2}}{\cancel{30}}} = 70 \text{ bu., } \textit{Ans.}$$

(25.) 325 lb. : 1625 lb. : : \$22.60 : ?

$$\frac{\overset{\overset{5}{\cancel{65}}}{\cancel{1625}} \times 22.60}{\underset{\cancel{13}}{\cancel{325}}} = \$113.00, \textit{ Ans.}$$

(26.) $4\frac{1}{2}$ ft. : 180 ft. : : 3 ft. : ?

$$180 \times 3 = 540; \quad \frac{\overset{60}{\cancel{540}}}{1} \times \frac{2}{\cancel{9}} = 120 \text{ ft., } \textit{Ans.}$$

(27.) 12 : 9 : : 60 : ?

$$\frac{9 \times \overset{5}{\cancel{60}}}{\cancel{12}} = 45 \text{ da., } \textit{Ans.}$$

(28.) 100 : 60 : : 2200 : ? $\frac{2200 \times 60}{100} = \$1320$, A's. } *Ans.*
100 : 60 : : 1800 : ? $\frac{1800 \times 60}{100} = \$1080$, B's. }

(29.) \$800.30 + \$250 + \$375.10 + \$500 + \$115 = \$2040.40. \$2040.40 : \$612.12 : : \$1.00 : ?
\$612.12 ÷ 2040.40 = \$0.30, *Ans.*

(30.) \$6 : \$8 : : 9 oz. : ?

$$\frac{\overset{4}{\cancel{8}} \times \overset{3}{\cancel{9}}}{\underset{2}{\cancel{6}}} = 12 \text{ oz., } \textit{Ans.}$$

(31.) \$300 : \$250 : : 6 mo. : ?

$$\frac{\overset{5}{\cancel{250}} \times \cancel{6}}{\underset{\cancel{6}}{\cancel{300}}} = 5 \text{ mo., } \textit{Ans.}$$

(32.) $27 \times 7 = 189$; $36 - 27 = 9$.
9 mi. : 189 mi. : : 1 da. : ? $\frac{189}{9} = 21$ da., *Ans.*

(33.) 9 hr. : 12 hr. : : \$$15\frac{2}{3}$: ? = \$$20.88\frac{8}{9}$, or 1 mo.'s services when he works 12 hr. a day.
\$$20.88\frac{8}{9} \times 4\frac{2}{5}$ = \$$91.91\frac{1}{9}$, *Ans.*

(34.) As 5 lb. : $\frac{3}{4}$ lb. : : \$$\frac{5}{8}$: \$$\frac{3}{32}$, *Ans.*

(35.) As 6 yd. : $7\frac{3}{8}$ yd. : : \$$5\frac{3}{5}$: \$$6\frac{53}{60}$, *Ans.*

(36.) As $\frac{1}{3}$ bu. : $\frac{1}{2}$ bu. : : \$$\frac{3}{8}$: \$$\frac{9}{16}$, *Ans.* ($\frac{3}{1} \times \frac{1}{2} \times \frac{3}{8} = \frac{9}{16}$.)

(37.) As $1\frac{3}{4}$ yd. : 2 yd. : : \$$\frac{7}{24}$: \$$\frac{1}{3}$, *Ans.* ($\frac{4}{7} \times \frac{2}{1} \times \frac{7}{24} = \frac{1}{3}$.)

(38.) As \$$29\frac{3}{4}$: \$$31\frac{1}{4}$: : $59\frac{1}{2}$ yd. : ? By cancellation, $\frac{4}{119} \times \frac{125}{4} \times \frac{119}{2} = \frac{125}{2} = 62\frac{1}{2}$ yd., *Ans.*

(39.) As .85 gal. : .25 gal. : : \$1.36 : \$0.40, *Ans.*

(40.) As 61.3 lb. : 1.08 lb. : : \$44.9942 : \$0.79, *Ans.*

(41.) As $\frac{5}{7}$ yd. : $\frac{9}{11}$ yd. : : \$$\frac{3}{5}$: \$$\frac{189}{275}$, *Ans.*

(42.) As $\frac{3}{7}$ yd. : $17\frac{3}{8}$ yd. : : \$$4\frac{2}{5}$: ?
$\frac{7}{3} \times \frac{139}{8} \times \frac{22}{5}$ = \$$178.38\frac{1}{3}$, *Ans.*

(43.) As 26 cogs : 35 cogs : : 1 rev. : $1\frac{9}{26}$ rev. Hence, the smaller wheel gains $\frac{9}{26}$ of a revolution in each revolution of the larger wheel. Then, $\frac{9}{26}$ rev : 10 rev. : : 1 rev. of larger : $28\frac{8}{9}$ revolutions of larger, *Ans.*

(44.) 1 gal. $=$ 32 gills; $32-1=31$. As 32 : 31 : : 100 gal. : $96\frac{7}{8}$ gal., *Ans.*

(45.) As 70 p. : 20 p. : : 60 sec. : $17\frac{1}{7}$ sec.
1142 ft. $\times 17\frac{1}{7}=19577\frac{1}{7}$ ft. $=$ 3 mi. 226 rd. 2 yd. $2\frac{1}{7}$ ft., *Ans.*

(46.) As 25 ft. : 25 ft. 5.25 in. : : 643 ft. 8 in. : 654 ft. 11.17 in., *Ans.*

Art. 225.

(3.) 2 da. : 10 da. }
4 hr. : 8 hr. } : : 24 mi. : 240 mi., *Ans.*

(4.) As 18 rd. : 72 rd. The more rods, the more men.
And as 8 da. : 12 da. The less days, the more men.
: : 16 men : 96 men, *Ans.*

(5.) As 6 p. : 15 p. The more persons, the more [dollars.
8 mo. : 20 mo. The more months, the more [dollars.
: : $150 : $937.50, *Ans.*

(6.) As 7 da. : 9 da. The more days, the more miles.
6 hr. : 11 hr. The more hours, the more mi.
: : 217 mi. : $511\frac{1}{2}$ mi., *Ans.*

(7.) As $100 : $75. The less dollars, the less interest.
12 mo. : 9 mo. The less months, the less interest.
: : $6 : $3.375, *Ans.*

(8.) As 10100 lb. : 100 lb. The more lb., the less miles.
20 ct. : $60.60 The more money, the more [miles.
: : 20 mi. : 60 mi., *Ans.*

(9.) As 12 cwt. 75 lb. : 10 T. The more weight, the more money.
400 mi. : 75 mi. The less miles, the less money.
: : $57.12 : $168, *Ans.*

(10.) As 20 men : 18 men. The more men, the less days.
40 rd. l. : 87 rd. l. The more length, the more days.
5 ft. h. : 8 ft. h. The more height, the more days.
4 ft. t. : 5 ft. t. The more thickness, the more days.
: : 15 days : $58\frac{29}{40}$ days, *Ans.*

(11.) As 100 men : 180 men. The less men, the more days.
200 yd. l. : 180 yd. l. The less length, the less days.
3 yd. w. : 4 yd. w. The more width, the more days.
2 yd. d. : 3 yd. d. The more depth, the more days.
8 hr. : 10 hr. The less hours, the more days.
: : 6 days : 24.3 days, *Ans.*

Art. 226.

(2.) $\frac{300}{800} = \frac{3}{8}$; $\frac{3}{8}$ of \$232 = \$87, A's share.
$\frac{500}{800} = \frac{5}{8}$; $\frac{5}{8}$ of \$232 = \$145, B's share.

(3.) \$70 + \$150 + \$80 = \$300, whole stock.
$\frac{70}{300} = \frac{7}{30}$; $\frac{7}{30}$ of \$120 = \$28, A's share.
$\frac{150}{300} = \frac{1}{2}$; $\frac{1}{2}$ of \$120 = \$60, B's share.
$\frac{80}{300} = \frac{4}{15}$; $\frac{4}{15}$ of \$120 = \$32, C's share.

(4.) \$200 + \$400 + \$600 = \$1200, whole stock. $\frac{200}{1200} = \frac{1}{6}$, $\frac{400}{1200} = \frac{1}{3}$, $\frac{600}{1200} = \frac{1}{2}$. $\frac{1}{6}$ of \$427.26 = \$71.21, A's share; $\frac{1}{3}$ of \$427.26 = \$142.42, B's share; and $\frac{1}{2}$ of \$427.26 = \$213.63, C's share.

(5.) $1+3+5=9$. $\frac{1}{9}$ of \$90 = \$10; $\frac{3}{9}=\frac{1}{3}$ of \$90 = \$30; $\frac{5}{9}$ of \$90 = \$50, *Ans.*

(6.) $2+3+5+7=17$. $\frac{2}{17}$ of \$735.93 = \$86.58; $\frac{3}{17}$ of \$735.93 = \$129.87; $\frac{5}{17}$ of \$735.93 = \$216.45; $\frac{7}{17}$ of \$735.93 = \$303.03, *Ans.*

(7.) $3+6+9+11+13+17=59$.
$\frac{3}{59}$ of \$22361 = \$1137; $\frac{6}{59}$ of \$22361 = \$2274;
$\frac{9}{59}$ of \$22361 = \$3411; $\frac{11}{59}$ of \$22361 = \$4169;
$\frac{13}{59}$ of \$22361 = \$4927; $\frac{17}{59}$ of \$22361 = \$6443, *Ans.*

(8.) $\frac{1}{3}, \frac{3}{5}, \frac{7}{8} = \frac{40}{120}, \frac{72}{120}, \frac{105}{120}$. Since the denominators are the same, the fractions are to each other as their numerators. $40+72+105=217$. $\frac{40}{217}$ of \$692.23 = \$127.60; $\frac{72}{217}$ of \$692.23 = \$229.68; $\frac{105}{217}$ of \$692.23 = \$334.95, *Ans.*

Art. 227.

(1.) \$175 + \$500 + \$600 + \$210 + \$42.50 + \$20 + \$10 = \$1557.50

As \$1557.50 : \$175 : : \$934.50 : \$105.00, A's share.
As \$1557.50 : \$500 : : \$934.50 : \$300.00, B's share.
As \$1557.50 : \$600 : : \$934.50 : \$360.00, C's share.
As \$1557.50 : \$210 : : \$934.50 : \$126.00, D's share.
As \$1557.50 : \$42.50 : : \$934.50 : \$25.50, E's share.
As \$1557.50 : \$20 : : \$934.50 : \$12.00, F's share.
As \$1557.50 : \$10 : : \$934.50 : \$6.00, G's share.

(2.) \$234 + \$175 + \$326 = \$735; \$492.45 ÷ 735 = \$0.67 = sum paid on each dollar of indebtedness. \$234 × .67 = \$156.78, A; \$175 × .67 = \$117.25, B; \$326 × .67 = \$218.42, C.

(3.) \$25000 — \$4650 = \$20350.
37000 : 20350 : : \$1 : \$0.55, *Ans.*

Art. 228.

(1.) $\frac{48}{108}=\frac{4}{9}$; $\frac{36}{108}=\frac{1}{3}$; $\frac{24}{108}=\frac{2}{9}$. $\frac{4}{9}$ of $45=20$, A's loss: $\frac{1}{3}$ of $45=15$, B's loss: $\frac{2}{9}$ of $45=10$, C's loss.

(2.) \$10000 + \$15000 = \$25000. 1125 ÷ 25000 = $.04\frac{1}{2}=4\frac{1}{2}\%$, gen. av. \$2150 × $.04\frac{1}{2}$ = \$96.75, A's loss.

Art. 229.

(3.) $23\times27=621$; $21\times39=819$; $621+819=1440$. $\frac{621}{1440}=\frac{69}{160}$; $\frac{819}{1440}=\frac{91}{160}$; $\frac{69}{160}$ of \$54 = \$23.28$\frac{3}{4}$, A pays; $\frac{91}{160}$ of \$54 = \$30.71$\frac{1}{4}$, B pays.

(4.) \$300 × 5 = \$1500; \$400 × 8 = \$3200; \$500 × 3 = \$1500. \$1500 + \$3200 + \$1500 = \$6200. $\frac{1500}{6200}=\frac{15}{62}$; $\frac{3200}{6200}=\frac{16}{31}$. $\frac{15}{62}$ of \$100 = \$24.19$\frac{11}{31}$, A's and C's loss; $\frac{16}{31}$ of \$100 = \$51.61$\frac{9}{31}$, B's loss.

(5.) $6\times30=180$: $5\times40=200$: $8\times28=224$. $180+200+224=604$; $\frac{180}{604}=\frac{45}{151}$; $\frac{200}{604}=\frac{50}{151}$; $\frac{224}{604}=\frac{56}{151}$. $\frac{45}{151}$ of \$18.12 = \$5.40, A: $\frac{50}{151}$ of \$18.12 = \$6, B: $\frac{56}{151}$ of \$18.12 = \$6.72, C.

(6.) A, \$300 × 8 = \$2400; \$300 + \$100 = \$400;
\$400 × 8 = \$3200. \$2400 + \$3200 = \$5600
B, \$600 × 10 = \$6000; \$600 − \$300 = \$300;
\$300 × 6 = \$1800. \$6000 + \$1800 = \$7800
\$13400

As \$13400 : \$5600 : : \$442.20 : \$184.80, A's.
\$13400 : \$7800 : : \$442.20 : \$257.40, B's.

(7.) \$800 × 12 = \$9600; \$500 × 12 = \$6000: 12 mo. — 7 mo. = 5 mo. \$9600 — \$6000 = \$3600; \$3600 ÷ 5 = \$720, *Ans.*

Art. 230.

(2)

\$2 × 4 = \$8
6 × 8 = 48
\$8) \$56(7 mo., *Ans.*

(3)

\$8 × 5 = \$40
4 × 8 = 32
\$12) \$72(6 mo., *Ans.*

(4)

$\$250 \times 2 = \$500$

$500 \times 5 = 2500$

$\underline{750 \times 8 = 6000}$

$\$1500) \quad \$9000(6$ mo., *Ans.*

(5)

$\$100 \times 6 = \$600$

$75 \times 8 = 600$

$\underline{125 \times 12 = 1500}$

$\$300) \quad \$2700(9$ mo., *Ans.*

(6)

$\frac{1}{5}$ of $\$200 = \$40$

$\frac{2}{5}$ of $\$200 = \$80$

$\$40 \times 0 = 0$

$80 \times 5 = \$400$

$\underline{80 \times 10 = 800}$

$\$200) \quad \$1200(6$ mo., *Ans.*

Art. 231.

(2.) Counting from April 2d, it is 90 days to the first payment, and 150 days to the second.

$\$200 \times 90 = \$18000$

$\underline{300 \times 150 = 45000}$

$\$500) \quad \$63000(126$ da. from April 2d = Aug. 6th, *Ans.*

(3.) Counting from July 6, when first bill is due,

$\$1250 \times 0 = 0$

$4280 \times 73 = 312440$

$\underline{675 \times 168 = 113400}$

$\$6205) \quad \$425840(68.6+$

Counting 69 days from July 6th, brings the time to Sept. 13, *Ans.*

Art. 232.

(2)

6 lb. at 3 ct. = 18 ct.

4 lb. at 8 ct. = 32 ct.

10 lb. cost 50 ct.

50 ct. ÷ 10 = 5 ct., *Ans.*

(3)

25 lb. at 12 ct. = $3.00

25 lb. at 18 ct. = 4.50

40 lb. at 25 ct. = 10.00

90 lb. cost $17.50

$\$17.50 \div 90 = \$0.19\frac{4}{9}$, *Ans.*

(4)

3 gal. cost	\$0.00
12 gal. at 50 ct. =	6.00
15 gal. cost	\$6.00

\$6.00 ÷ 15 = \$0.40, *Ans.*

(5)

10 at \$3 =	\$30.00
12 at 4 =	48.00
8 at 9 =	72.00
30 worth	\$150.00

\$150 ÷ 30 = \$5, *Ans.*

(6)

6 to 10 = 4 hr.	63° × 4 = 252°
10 to 1 = 3 "	70° × 3 = 210°
1 to 3 = 2 "	75° × 2 = 150°
3 to 7 = 4 "	73° × 4 = 292°
7 to 6 = 11 "	55° × 11 = 605°
24)	1509°(62$\frac{7}{8}$°, *Ans.*

INVOLUTION.

Art. 234.

(2.) 65 × 65 = 4225, *Ans.*

(3.) 25 × 25 × 25 = 15625, *Ans.*

(4.) 12 × 12 × 12 × 12 = 20736, *Ans.*

(5.) 10 × 10 × 10 × 10 × 10 = 100000, *Ans.*

(6.) 9 × 9 × 9 × 9 × 9 × 9 = 531441, *Ans.*

(7.) 2 × 2 × 2 × 2 × 2 × 2 × 2 × 2 = 256, *Ans.*

(8.) $\frac{2}{3} \times \frac{2}{3} = \frac{4}{9}$, *Ans.* (9.) $\frac{3}{4} \times \frac{3}{4} \times \frac{3}{4} = \frac{27}{64}$, *Ans.*

(10.) $\frac{4}{5} \times \frac{4}{5} \times \frac{4}{5} \times \frac{4}{5} = \frac{256}{625}$, *Ans.*

(11.) $\frac{2}{3} \times \frac{2}{3} \times \frac{2}{3} \times \frac{2}{3} \times \frac{2}{3} = \frac{32}{243}$, *Ans.*

(12.) $16\frac{1}{2} = \frac{33}{2}$. $\frac{33}{2} \times \frac{33}{2} = \frac{1089}{4} = 272\frac{1}{4}$, *Ans.*

(13.) $12\frac{1}{2} = \frac{25}{2}$. $\frac{25}{2} \times \frac{25}{2} \times \frac{25}{2} = \frac{15625}{8} = 1953\frac{1}{8}$, *Ans.*

(14.) .25 × .25 × .25 × .25 = .00390625, *Ans.*

(15.) 14 × 14 × 14 = 2744, *Ans.*

(16.) $19 \times 19 \times 19 \times 19 = 130321$, *Ans.*

(17.) $2\frac{1}{3} = \frac{7}{3}$. $\frac{7}{3} \times \frac{7}{3} \times \frac{7}{3} \times \frac{7}{3} \times \frac{7}{3} = \frac{16807}{243} = 69\frac{40}{243}$, *Ans*

Art. 238. EVOLUTION.

(5)

529(20 + 3 = 23, *Ans.*
400
20 | 129
2
40
3
43 | 129

(6)

625(25, *Ans.*
4
45)225
225

(7)

6561(81, *Ans.*
64
161)161
161

(8)

56644(238, *Ans.*
4
43)166
129
468)3744
3744

(9)

390625(625, *Ans.*
36
122)306
244
1245)6225
6225

(10)

1679616(1296, *Ans.*
1
22)67
44
249)2396
2241
2586)15516
15516

(11)

5764801(2401, *Ans.*
4
44)176
176
4801)4801
4801

(12)

43046721(6561, *Ans.*
36
125)704
625
1306)7967
7836
13121)13121
13121

(13)

987656329(31427, *Ans.*
9
61)87
61
624)2665
2496
6282)16963
12564
62847)439929
439929

(14)

```
 . . . . .
 289442169(17013,
 1              Ans.
27)189
   189
 3401)4421
      3401
34023)102069
      102069
```

(15)

```
  . . .
  234.09(15.3,
  1        Ans.
25)134
   125
 303)909
     909
```

(16)

```
   . . . . .
   145.2025(12.05,
   1           Ans.
 22)45
    44
2405)12025
     12025
```

(17)

```
 . . . .
 915.0625(30.25, Ans.
 9
692)1506
    1204
6045)30225
     30225
```

(18)

```
 . .
 .0196(.14, Ans.
   1
 24)96
    96
```

(19)

```
 . . . .
 1.008016(1.004, Ans.
 1
2004)008016
       8016
```

(20)

```
   . . .
 .00822649(.0907, Ans.
    81
1807)12649
     12649
```

(21.) $\sqrt{25}=5$, $\sqrt{729}=27$; $\sqrt{\frac{25}{729}}=\frac{5}{27}$, *Ans.*

(22.) $\frac{847}{1183}=\frac{121}{169}$; $\sqrt{121}=11$, $\sqrt{169}=13$; *Ans.* $=\frac{11}{13}$.

(23.) $30\frac{1}{4}=\frac{121}{4}$; $\sqrt{\frac{121}{4}}=\frac{11}{2}=5\frac{1}{2}$, *Ans.*

(24)

10(3.162277+, *Ans.*
9
61)100
61
626)3900
3756
6322)14400
12644
63242)175600
126484
632447)4911600
4427129
6324547)48447100
44271829

(25)

2(1.41421+, *Ans.*
1
24)100
96
281)400
281
2824)11900
11296
28282)60400
56564
282841)383600
282841

(27)

$6\frac{2}{5}$ = 6.4(2.5298+, *Ans.*
4
45)240
225
502)1500
1004
5049)49600
45441
50588)415900
404704

(28)

$384\frac{4}{7}$ = 384.5714285714(19.61049+, *Ans.*
1
29)284
261
386)2357
2316
3921)4114
3921
392204)1932857
1568816
3922089)36404114
35298801

(26)

$\frac{2}{3}$ = .666666+(.81649+, *Ans.*
64
161)266
161
1626)10566
9756
16324)81066
65296
163289)1577066
1469601

Art. 239.

(2.) $16 = 2 \times 2 \times 2 \times 2$: $\sqrt{16} = 2 \times 2 =$ 4, *Ans.*

(3.) $36 = 2 \times 2 \times 3 \times 3$: $\sqrt{36} = 2 \times 3 =$ 6, *Ans.*

(4.) $100 = 2 \times 2 \times 5 \times 5$: $\sqrt{100} = 2 \times 5 = 10$, *Ans.*

(5.) $225 = 3 \times 3 \times 5 \times 5$: $\sqrt{225} = 3 \times 5 = 15$, *Ans.*

(6.) $\sqrt{(16 \times 25)} = 4 \times 5 = 20$, *Ans.*

(7.) $\sqrt{(36 \times 49)} = 6 \times 7 = 42$, *Ans.*

(8.) $\sqrt{(64 \times 81)} = 8 \times 9 = 72$, *Ans.*

(9.) $\sqrt{(121 \times 25)} = 11 \times 5 = 55$, *Ans.*

Art. 240.

(1)	(2)	(3)
$30^2 = 900$	$100^2 = 10000$	$45^2 = 2025$
$40^2 = 1600$	$60^2 = 3600$	$60^2 = 3600$
$\sqrt{2500} = 50$, *Ans.*	$\sqrt{6400} = 80$, *Ans.*	$\sqrt{5625} = 75$, *Ans.*

(4.) $60^2 = 3600$, $37^2 = 1369$; $3600 - 1369 = 2231$; $\sqrt{2231} = 47.2334+ =$ width of street from foot of ladder on one side. $60^2 = 3600$, $23^2 = 529$; $3600 - 529 = 3071$; $\sqrt{3071} = 55.4166+ =$ width of street from foot of ladder on the other side. $47.2344 + 55.4166 = 102.65$, *Ans.*

(5.) $600^2 = 360000$, $140^2 = 19600$; $360000 - 19600 = 340400$; $\sqrt{340400} = 583.43+$; $100 \div 2 = 50$; $583.43 - 50. = 533.43+$, *Ans.*

(6)

$20^2 = 400$
$16^2 = 256$
656
Square of base $= 656$

The square root of 656, will give the length of the diagonal line joining opposite corners of the floor of the room: this is the base of the triangle, of which the hypotenuse is required.

$12^2 =$ perpendicular$^2 = 144$; $656 + 144 = 800$; $\sqrt{800} = 28.28+$, *Ans.*

Art. 241.

(1)

6241(79 rd., *Ans.*
49
149)1341
1341

(2)

8 sq. ft. 4 sq. in. = 1156 sq. in.
1156(34 in. = 2 ft. 10 in., *Ans.*
9
64) 256
256

(3.) $\sqrt{4096} = 64$ yd., *Ans.*

(4.) $4 \times 4 = 16$; $16 \times 9 = 144$; $\sqrt{144} = 12$, *Ans.*
Or, $\sqrt{(16 \times 9)} = 4 \times 3 = 12$ rd., *Ans.*

(5.) There are 43560 sq. ft. in 1 acre.
$\sqrt{43560} = 208.71+$ ft., side of acre.

Art. 244.

(3)

2197(13
1
300 | 1197
90
9
399 | 1197

13824(24
8
1200 | 5824
240
16
1456 | 5824

$\frac{13}{24}$, *Ans.*

(4)

.800000000(.928, *Ans.*
729
24300 | 71000
540
4
24844 | 49688
2539200 | 21312000
22080
64
2561344 | 20490752

(5)

$9\dot{1}12\dot{5}$(45, *Ans.*

	64
	27125
$4\times4\times300=4800$	
$4\times5\times\ 30=\ 600$	
$5\times5\quad=\ 25$	
5425	27125

(6)

$19\dot{5}11\dot{2}$(58, *Ans.*

	125
	70112
$5\times5\times300=7500$	
$5\times8\times\ 30=1200$	
$8\times8\quad=\ 64$	
8764	70112

(7)

$91\dot{2}67\dot{3}$(97, *Ans.*

	729
	183673
24300	
1890	
49	
26239	183673

(8)

$\dot{1}225043$(167, *Ans.*

	1
$1\times\ 1\times300=\ 300$	225
$10\times10\times300=30000$	225043
$10\times\ 7\times\ 30=\ 2100$	
$7\times\ 7\quad=\ 49$	
32149	225043

(9)

$1\dot{3}31\dot{2}05\dot{3}$(237, *Ans.*

	8
$2\times2\times300=1200$	5312
$2\times3\times\ 30=\ 180$	
$3\times3\quad=\ 9$	
1389	4167
$23\times23\times300=158700$	1145053
$23\times\ 7\times\ 30=\ 4830$	
$7\times\ 7\quad=\ 49$	
163579	1145053

(10)

	102503232(468, *Ans.*
	64
4×4×300=4800	38503
4×6× 30= 720	
6×6 = 36	
5556	33336
46×46×300=634800	5167232
46× 8× 30= 11040	
8× 8 = 64	
645904	5167232

(11)

	529475129(809, *Ans.*
	512
8× 8×300= 19200	17475
80×80×300=1920000	17475129
80× 9× 30= 21600	
9× 9 = 81	
1941681	17475129

(12)

	958585256(986, *Ans.*
	729
9×9×300=24300	229585
9×8× 30= 2160	
8×8 = 64	
26524	212192
98×98×300=2881200	17393256
98× 6× 30= 17640	
6× 6 = 36	
2898876	17393256

(13)

14760213677(2453, *Ans.*

	8
$2\times 2\times 300=1200$	6760
$2\times 4\times 30=240$	
$4\times 4=16$	
1456	5824
$24\times 24\times 300=172800$	936213
$24\times 5\times 30=3600$	
$5\times 5=25$	
176425	882125
$245\times 245\times 300=18007500$	54088677
$245\times 3\times 30=22050$	
$3\times 3=9$	
18029559	54088677

(14)

128100283921(5041, *Ans.*

	125
$5\times 5\times 300=7500$	3100
$50\times 50\times 300=750000$	3100283
$50\times 4\times 30=6000$	
$4\times 4=16$	
756016	3024064
$504\times 504\times 300=76204800$	76219921
$504\times 1\times 30=15120$	
$1\times 1=1$	
76219921	76219921

(15) 53.157376(3.76, *Ans.*

	27
$3\times3\times300=2700$	26157
$3\times7\times\ 30=\ \ 630$	
$7\times7\quad=\ \ 49$	
3379	23653
$37\times37\times300=410700$	2504376
$37\times\ 6\times\ 30=\ \ 6660$	
$6\times\ 6\quad=\ \ 36$	
417396	2504376

(16) .199176704(.584, *Ans.*

	125
$5\times5\times300=7500$	74176
$5\times8\times\ 30=1200$	
$8\times8\quad=\ \ 64$	
8764	70112
$58\times58\times300=1009200$	4064704
$58\times\ 4\times\ 30=\ \ 6960$	
$4\times\ 4\quad=\ \ 16$	
1016176	4064704

(17.) $\sqrt[3]{216}=6.$ $\sqrt[3]{343}=7.$ *Ans.* $\frac{6}{7}$.

(18.) $\sqrt[3]{2744}=14.$ $\sqrt[3]{6859}=19.$ *Ans.* $\frac{14}{19}$.

(19.) $\frac{48778}{118638}=\frac{24389}{59319}$ $\sqrt[3]{24389}=29.$ $\sqrt[3]{59319}=39.$ *Ans.* $\frac{29}{39}$.

(20.) $5\frac{104}{125}=\frac{729}{125}$ $\sqrt[3]{729}=9.$ $\sqrt[3]{125}=5.$ *Ans.* $\frac{9}{5}=1\frac{4}{5}$.

(21)	2(1.259+, *Ans.*
	1
	1000
300+60+4=364	728
12×12×300=43200	272000
12× 5× 30= 1800	
5× 5 = 25	
45025	225125
125×125×300=4687500	46875000
125× 9× 30= 33750	
9× 9 = 81	
4721331	42491979

(22)	9(2.080+, *Ans.*
	8
2×2×300=1200	1000
20×20×300=120000	1000000
20× 8× 30= 4800	
8× 8 = 64	
124864	998912
208×208×300=12979200	1088000

(23)	200(5.848+, *Ans.*
	125
5×5×300=7500	75000
5×8× 30=1200	
8× 8= 64	
8764	70112
58×58×300=1009200	4888000
58× 4× 30= 6960	
4× 4= 16	
1016176	4064704
584×584×300=102316800	823296000
584× 8× 30= 140160	
8× 8= 64	819656192

(24)

$9\frac{1}{6} = 9.166666+(2.092+$, *Ans.*

	8
2×2×300=1200	1166
20×20×300=120000	1166666
20× 9× 30= 5400	
9× 9= 81	
125481	1129329
209×209×300=13104300	37337666
209× 2× 30= 12540	
2× 2= 4	
13116844	26233688

Art. 245.

(1.) $\sqrt[3]{1953.125} = 12.5$ ft., *Ans.*

(2.) $64 \times 3 \times 3 \times 3 = 1728$ cu. in. = 1 cu. ft., one side of which = 1 ft., *Ans.*

(3.) $\sqrt[3]{512} = 8$ half in. = 4 in., *Ans.*

(4.) 450 cu. yd. 17 cu. ft. = 12167 cu. ft.; $\sqrt[3]{12167} =$ 23 ft., *Ans.*

(5.) $288 \times 216 \times 48 = 2985984$, $\sqrt[3]{2985984} = 144$ ft., *Ans.*

(6.) $1728 \times 3 = 5184$, $\sqrt[3]{5184} = 17.306+$ in., *Ans.*

MENSURATION.

Art. 247.

(1.) 17 ft. × 15 ft. = 255 sq. ft., *Ans.*

(2.) 120 rd. × 84 rd. = 10080 sq. rd. = 63 A., *Ans.*

(3.) 65 rd. × 65 rd. = 4225 sq. rd. = 26 A. 65 sq. rd.

(4.) 35 rd. × 16 rd. = 560 sq. rd. = 3 A. 80 sq. rd., *Ans.*

(5.) 30 ft. × 30 ft. = 900 sq. ft. = 100 sq. yd.
15 ft. × 15 ft. = 225 sq. ft., × 2 = 450 sq. ft. = 50 sq. yd.
Diff. = 50 sq. yd.

(7.) 5 ft. 6 in. = $5\frac{1}{2}$ ft.; 1 ft. 8 in. = $1\frac{2}{3}$ ft.; $\frac{11}{2}$ ft. × $\frac{5}{3}$ ft. = $\frac{55}{6}$ sq. ft. = $9\frac{1}{6}$ sq. ft., *Ans.*

(8.) 25 ft. 9 in. = $25\frac{3}{4}$ ft. = $\frac{103}{4}$ ft.; 21 ft. 3 in. = $21\frac{1}{4}$ ft. = $\frac{85}{4}$ ft.; $\frac{103}{4} \times \frac{85}{4} = \frac{8755}{16}$ sq. ft. = $547\frac{3}{16}$ sq. ft. = 60 sq. yd. 7 sq. ft. 27 sq. in., *Ans.*

(9.) 80 sq. ft. ÷ 10 ft. = 8 ft., *Ans.*

(10.) 18 ft. × 15 ft. = 270 sq. ft. = 30 sq. yd.; 30 sq. yd. ÷ $1\frac{1}{2}$ yd. = 20 yd., *Ans.*

(11.) 3 yd. × $1\frac{1}{2}$ yd. = $4\frac{1}{2}$ sq. yd.; $4\frac{1}{2} \div \frac{3}{4} = \frac{9}{2} \times \frac{4}{3} = 6$ yd., *Ans.*

(12.) 21 ft. 3 in. = 21.25 ft.; 13 ft. 6 in. = 13.5 ft.; 21.25 ft. × 13.5 ft. = 286.875 sq. ft.; $1\frac{1}{4}$ yd. = $3\frac{3}{4}$ = 3.75 ft.; 286.875 sq. ft. ÷ 3.75 ft. = 76.5 ft. = 25.5 yd. = $25\frac{1}{2}$ yd.

(13.) 160 sq. rd. in 1 A. 160 ÷ 15 = $10\frac{2}{3}$ rd., *Ans.*

Art. 248.

(1.)

ft.	in.
2	2
2	11
5	1 = 61 in.

61 in. ÷ 2 = $30\frac{1}{2}$ in., × 11 in. = $335\frac{1}{2}$ sq. in.: $335\frac{1}{2}$ sq. in. ÷ 144 = 2 sq. ft. $47\frac{1}{2}$ sq. in., *Ans.*

(2.) 25 rd. + 19 rd. = 44 rd.; 44 rd. ÷ 2 = 22 rd., × 32 rd. = 704 sq. rd., ÷ 160 = 4 A. 64 sq. rd., *Ans.*

(3.) 10 ft. 8 in. = 128 in.; 6 ft. 2 in. = 74 in.; 128 + 74 = 202 in., ÷ 2 = 101 in.; 12 ft. = 144 in.; 101 × 144 = 14544 sq. in. = 101 sq. ft. = 11 sq. yd. 2 sq. ft., *Ans.*

Art. 249.

(1.) 15 ft. × 12 ft. = 180 sq. ft., ÷ 2 = 90 sq. ft., *Ans.*

(2.) 44 rd. $\times$ 18 rd. $=$ 792 sq. rd., $\div$ 2 $=$ 396 sq. rd.: 396 sq. rd. $\div$ 160 $=$ 2 A. 76 sq. rd., *Ans.*

(3.) $12\frac{1}{2}$ ft. $\times$ $16\frac{3}{4}$ ft. $= \frac{1675}{8}$ sq. ft. $= 209\frac{3}{8}$ sq. ft.; $209\frac{3}{8}$ $\div$ 2 $= 104\frac{11}{16}$ sq. ft. $=$ 11 sq. yd. 5 sq. ft. 99 sq. in., *Ans.*

(4.) $13 + 14 + 15 = 42, \div 2 = 21$. $21 - 13 = 8$, $21 - 14 = 7$, $21 - 15 = 6$. $21 \times 8 \times 7 \times 6 = 7056$: its square root $=$ 84 sq. ft., *Ans.*

(5.) $30 + 40 + 50 = 120, \div 2 = 60$. $60 - 30 = 30$, $60 - 40 = 20$, $60 - 50 = 10$. $60 \times 30 \times 20 \times 10 = 360000$: $\sqrt{360000} = 600$ sq. ft.: 600 sq. ft. $=$ 66 sq. yd. 6 sq. ft., *Ans.*

Art. 250.

(1.) 50 rd. $\times$ 30 rd. $=$ 1500 sq. rd., $\div$ 2 $=$ 750 sq. rd.: 50 rd. $\times$ 20 rd. $=$ 1000 sq. rd., $\div$ 2 $=$ 500 sq. rd.; 750 sq. rd. $+$ 500 sq. rd. $=$ 1250 sq. rd. $=$ 7 A. 130 sq. rd., *Ans.*

Art. 251.

(1.) 48 ft. $\times$ 3.1416 $=$ 150.7968 ft. $=$ 150 ft. 9.56 in.

(2.) 15 ft. $\div$ 3.1416 $=$ 4.7746 ft. $=$ 4 ft. 9.3 in. nearly.

(3.) 4 $\times$ 3.1416 $=$ 12.5664 ft. $=$ 12 ft. 6.8 in. nearly.

(4.) 12 ft. 5 in. $=$ 12.4166+ ft.; 12.4166 ft. $\div$ 3.1416 $=$ 3.952338 ft. $=$ 3 ft. 11.43 in. nearly, *Ans.*

(5.) 7912 mi. $\times$ 3.1416 $=$ 24856+ mi., *Ans.*

Art. 252.

(1.) $21 \times 21 = 441$: $3.1416 \times 441 = 1385.4456$ sq. ft. $=$ 153 sq. yd. 8 sq. ft. 64 sq. in., *Ans.*

NOTE.—To find the diameter when the area is given, divide the area by .7854; the square root of the quotient will be the diameter.

(2.) 6 sq. ft. 98.115 sq. in. $=$ 962.115 sq. in.; 962.115 $\div$ .7854 $=$ 1225: $\sqrt{1225} = 35$ in., $=$ 2 ft. 11 in. $=$ diameter. 35 in. $\times$ 3.1416 $=$ 109.956 in. $=$ 9 ft. 1.9+ in. $=$ circum.

(3.) 160 rd. $\div$ .7854 = 203.71785077+ ; $\sqrt{203.71785077}$ = 14.2729 ; 14.2729 $\div$ 2 = 7.1364 rd. = 7 rd. 2 ft. 3 in., *Ans.*

(4.) 10 $\div$ 2 = 5 = one radius ; $5^2 = 25$: 16 $\div$ 2 = 8 = one radius ; $8^2 = 64$. 25 $\times$ 3.1416 = 78.5400 ; 64 $\times$ 3.1416 = 201.0624 : 201.0624 — 78.5400 = 122.5224 sq. ft. ; .5224 $\times$ 144 = 75 sq. in. *Ans.* 122 sq. ft. 75 sq. in.

(5.) 1 sq. ft. = 144 sq. in. 144 $\div$ .7854 = 183.3460 sq. in.: $\sqrt{183.3460}$ = 13.54 in., *Ans.*

Art. 254.

(1.) 37 $\times$ 37 $\times$ 6 = 8214 sq. in. = 6 sq. yd. 3 sq. ft. 6 sq. in., *Ans.*

(2.) 4 + 4 + 4 = 12 ft., $\times$ 5 ft. = 60 sq. ft. = convex surface.

$$\frac{4+4+4}{2} = 6. \quad \left.\begin{matrix} 6-4=2 \\ 6-4=2 \\ 6-4=2 \end{matrix}\right\} 6 \times 2 \times 2 \times 2 = 48.$$

$\sqrt{48}$ = 6.92+ ; 6.92+. $\times$ 2 = 13.85 sq. ft. = area of 2 bases. 60 sq. ft. + 13.85 sq. ft. = 73.85+ sq. ft., *Ans.*

$$\text{(3.)}\ \left.\begin{matrix} 3 \text{ ft. } 6 \text{ in.} = 3\tfrac{1}{2} \text{ ft.} = \tfrac{7}{2}\,;\ \tfrac{7}{2} \times 2 = 7 \\ 2 \text{ ft. } 9 \text{ in.} = 2\tfrac{3}{4} \text{ ft.} = \tfrac{11}{4};\ \tfrac{11}{4} \times 2 = 5\tfrac{1}{2} \end{matrix}\right\} = 12\tfrac{1}{2} \text{ or } \tfrac{25}{2}.$$

$$1 \text{ ft. } 10 \text{ in.} = 1\tfrac{5}{6} \text{ ft.} = \tfrac{11}{6}.$$

$\frac{25}{2} \times \frac{11}{6} = \frac{275}{12}$ = convex surface ; $\frac{7}{2} \times \frac{11}{4} \times 2 = \frac{77}{4}$ or $\frac{231}{12}$ = areas of 2 bases : $\frac{275}{12} + \frac{231}{12} = \frac{506}{12} = 42\frac{1}{6}$ sq. ft., *Ans.*

(4.) 3.1416 $\times$ 4 ft. (diameter) = 12.5664 = circumference.

12.5664 $\times$ 5 = 62.8320 = convex surface.

2 $\times$ 2 $\times$ 3.1416 $\times$ 2 = 25.1328 = areas of 2 bases.

87.96+ sq. ft., *Ans.*

Art. 255.

(2.) 24 ft. $\times$ $18\frac{1}{2}$ ft. $\times$ $10\frac{7}{12}$ ft. = 4699 cu. ft. = 174 cu. yd. 1 cu. ft., *Ans.*

(3.) Area of base $= 1.73+$ sq. ft.; 1.73 sq. ft. $\times$ 14 ft. $= 24\frac{1}{4}$ cu. ft. nearly, *Ans.*

(4.) $2 \times 2 \times 3.1416 \times 12 = 150.8$ cu. ft., *Ans.*

(5.) $9\frac{1}{4}$ in. $= \frac{37}{4}$; $\frac{1}{2}$ of $\frac{37}{4} = \frac{37}{8}$; $(\frac{37}{8})^2 \times 3.1416 \times 8 = 537.6$ cu. in., *Ans.*

Art. 256.

(1.) 5 ft. 4 in. $= 5\frac{1}{3}$ ft.; $5\frac{1}{3}$ ft. $\times 3 = 16$ ft. $=$ perimeter of base. $7\frac{1}{2}$ ft. $\times 16 = 120$ sq. ft.; 120 sq. ft. $\div 2 = 60$ sq. ft. $=$ area of 3 sides. $5\frac{1}{3} \times 3 = 16$. $16 \div 2 = 8$; $8 - 5\frac{1}{3} = 2\frac{2}{3}$; $2\frac{2}{3} = \frac{8}{3}$; $8 \times \frac{8}{3} \times \frac{8}{3} \times \frac{8}{3} = \frac{4096}{27} = 151.70+$. $\sqrt{151.70} = 12.3+$ sq. ft. $=$ area of base. 60 sq. ft. $+ 12.3+$ sq. ft. $= 72.3+$ sq. ft., *Ans.*

(2.) $8\frac{1}{2}$ ft. $\times 3.1416 = 26.7036$ ft. $=$ circum. of base. $26.7036 \times 25 \div 2 = 333.79+$, *Ans.*

(3.) $2\frac{11}{12}$ ft. $\times 3.1416 \times 4\frac{7}{12}$ ft. $\div 2 = 21.008$ sq. ft. $=$ convex surface. $2\frac{11}{12}$ ft. $= \frac{35}{12}, \div 2 = \frac{35}{24}$; $(\frac{35}{24})^2 \times 3.1416 = 6.68$ sq. ft. $=$ area of base. $21.008 + 6.68 = 27.6+$ sq. ft., *Ans.*

Art. 257.

(1.) 5 ft. $\times$ 5 ft. $= 25$ sq. ft. $=$ area of base. 25 sq. ft. $\times$ 21 ft. $\div 3 = 175$ cu. ft., *Ans.*

(2.) $(5)^2 \times 3.1416 \times 15 \div 3 = 392.7$ cu. ft., *Ans.*

(3.) 720 ft. $= 240$ yd.; 477 ft. $= 159$ yd.; $(240 \text{ yd})^2 \times 159$ yd. $\div 3 = 3052800$ cu. yd., *Ans.*

(4.) $37\frac{2}{3}$ ft. $= \frac{113}{3}, \div 2 = \frac{113}{6}$; $(\frac{113}{6})^2 \times 3.1416 = 1114.3+$ sq. ft. $=$ area of base. $1114.3+$ sq. ft. $\times 79\frac{3}{4}$ ft., $\div 3 = 29622+$ cu. ft., *Ans.*

Art. 258.

(2.) $(4\frac{1}{2} \text{ ft.})^2 \times 3.1416 = 63.6+$ sq. ft., *Ans.*

(3.) $(7912)^2 \times 3.1416 = 196663355.75$ sq. mi., *Ans.*

Art. 259.

(1.) $13 \times 13 \times 13 \times .5236 = 1150.3+$ cu. ft., *Ans.*

(2.) $2\frac{1}{2}$ ft. $= \frac{5}{2}$: $\frac{5}{2} \times \frac{5}{2} \times \frac{5}{2} \times .5236 = 8.18+$ cu. ft., *Ans.*

(3.) 1 cu. ft. = 1728 cu. in.; $1728 \div .5236 = 3300.229$; $\sqrt[3]{3300.229} = 14.9$ in. nearly, *Ans.*

Art. 260.

(1.) $20\frac{1}{2} \times 16\frac{1}{4}$ = area of ceiling; $20\frac{1}{2} \times 10\frac{1}{12} \times 2$ = area of 2 sides; $16\frac{1}{4} \times 10\frac{1}{12} \times 2$ = area of other 2 sides. Add these amounts together, and deduct $6\frac{1}{4} \times 4\frac{1}{6}$, fire-place; $7 \times 4\frac{1}{6}$, door; $6 \times 3\frac{1}{4} \times 2$, two windows.

(2.) $20 \times 10\frac{1}{3} \times 2$ = area of two sides; $14\frac{1}{2} \times 10\frac{1}{3} \times 2$ = area of other 2 sides. Deduct $4 \times 4\frac{1}{3}$, fire-place; $6 \times 3\frac{1}{6} \times 2$, two windows. The remainder is in sq. ft. Divide by 9, and multiply by 27 ct. = \$19.73+, *Ans.*

(3.)
21 yd. × 15 yd. = 315 sq. yd.
5 ft. = $1\frac{2}{3}$ yd. 21 yd. × $1\frac{2}{3}$ yd. = 35 sq. yd.
280 sq. yd.

$35 \times .36 = \$12.60$; $280 \times .24 = \$67.20$; $\$12.60 + \$67.20 = \$79.80$, *Ans.*

(4.) $15\frac{1}{2}$ ft. $\times 12\frac{1}{2}$ ft. $\times 2 = 387.5$ sq. ft. = 43.06 sq. yd.; 43.06 sq. yd. @ 10 ct. = \$4.31, *Ans.*

(5.) 6 ft. 11 in. + 5 ft. 4 in. + 4 ft. 3 in., $\times 7 = 115\frac{1}{2}$ ft., $\times 3\frac{1}{2}$ ft. $= 404\frac{1}{4}$ sq. ft., × 16 ct. = \$64.68, *Ans.*

(6.) $36\frac{1}{4}$ ft. $\times 16\frac{1}{2}$ ft. = 598 sq. ft. = 5.98 squares; 5.98 × \$3.00 = \$17.94, *Ans.*

(7.) 40 ft. $\times 18\frac{1}{2}$ ft. $\times 2 = 1480$ sq. ft. = 14.80 squares; 14.80 × \$3.50 = \$51.80, *Ans.*

Art. 261.

(1.) $16 \times 1\frac{1}{4} = 20$ ft., *Ans.*

(2.) $12\frac{1}{2} \times 2\frac{1}{4} \times 2 = 56\frac{1}{4}$ ft., *Ans.*

(3.) $15 \times \frac{1}{3} \times 3 = 15$ ft., *Ans.*

(4.) $12 \times 2 \times 24 = 576$ ft., *Ans.*

(5.) 1 ft. 3 in. + 11 in. $= 2\frac{1}{6}$ ft., $\div 2 = 1\frac{1}{12}$ ft. = average width. $12\frac{1}{2} \times 1\frac{1}{12} = 13\frac{13}{24}$ ft., *Ans.*

Art. 262.

(1.) 97 ft. 5 in. = 97.416+ ft.; 18 ft. 3 in. = 18.25 ft.; 2 ft. 3 in. = 2.25 ft.: 97.416 ft. × 18.25 ft. × 2.25 ft. = 4000.1445 cu. ft. 4000.1445 ÷ 24.75 = 161.6+ P., *Ans.*

(2.) 53 ft. 6 in. = 53.5 ft.; 12 ft. 6 in. = 12.5 ft.; 53.5 × 12.5 × 2 = 1337.5 cu. ft. = 54.0404+ P. 54.0404 × \$2.25 = \$121.59+, *Ans.*

(3.) $48\frac{1}{3} \times 16\frac{1}{2} \times 1\frac{1}{2} = \frac{145}{3} \times \frac{33}{2} \times \frac{3}{2} = \frac{14355}{12} = 1196\frac{1}{4}$ cu. ft.; $1196\frac{1}{4} \times 20 = 23925$ bricks, *Ans.*

(4.) $120 \times 8 \times 1\frac{1}{2} = 1440$ cu. ft. = 2488320 cu. in. in wall; 8 × 4 × 2.25 = 72 cu. in. in each brick; 2488320 ÷ 72 = 34560 bricks, *Ans.*

(5.) 240 × 6 × 3 = 4320 cu. ft. = 7464960 cu. in. in wall; 9 × 4 × 2 = 72 cu. in. in brick; 7464960 ÷ 72 = 103680 bricks; 103680 ÷ 1000 = 103.68; \$3.25 × 103.68 = \$336.96, *Ans.*

Art. 263.

(1.) 15 ft. × 5 ft. × 4 ft. = 300 cu. ft. = 518400 cu. in.; 518400 ÷ 2150.4 = 241+ bu., *Ans.*

(2.) 10 ft. = 120 in.; 5 ft. = 60 in.; 4 ft. = 48 in.; 120 × 60 × 48 = 345600 cu. in.; 345600 ÷ 231 = 1496+ gal., *Ans.*

(3.) $(6)^2 \times .7854 = 28.2744$ = area of end; 28.2744 × 8 = 226.1952 cu. ft. = 390865.3056 cu. in. This divided by 2150.4 = 181.76+ bu., *Ans.*

(4.) 4 ft. $=48$ in.; 6 ft. $=72$ in.; $(48)^2=2304$; $2304 \times .7854 \times 72=130288.4352$ cu. in.; $130288.4352 \div 231 = 564.019+$ gal.; $564.019+ \div 31\frac{1}{2}=17.9+$ bl., *Ans.*

PROGRESSIONS.

Arithmetical Progression.

Case I.

Art. 265.

(3.) $50-1=49$; $49 \times 3+2=149$, *Ans.*

(4.) $54-1=53$; $53 \times 2=106$; $140-106=34$, *Ans.*

(5.) $99-1=98$; $98 \times \frac{7}{8}=85\frac{3}{4}$; $329-85\frac{3}{4}=243\frac{1}{4}$, *Ans.*

Case II.

Art. 266.

(2.) $300-3=297$; $10-1=9$; $297 \div 9=33$, *Ans.*

(3.) $50-5=45$; $10-1=9$; $45 \div 9=5$ miles, *Ans.*

Case III.

Art. 267.

(2.) $50+2=52$; $52 \times 24=1248$; $1248 \div 2=624$, *Ans.*

(3.) $1+12=13$; $13 \times 12=156$; $156 \div 2=78$ strokes, *Ans.*

(4.) The number of terms is evidently 100. The boy travels 6 yards to put the first apple in the basket, 12 the second, and so on; hence, the first term is 6, and the common difference 6. $100-1=99$; $99 \times 6+6= 600$, last term. $6+600=606$; $606 \times 100=60600$; $60600 \div 2=30300$ yd.; 30300 yd. $=17$ mi. 69 rd. $\frac{1}{2}$ yd., *Ans.*

(5.) Common difference $= 193 \times 2 = 386$ in. $60 - 1 = 59$; $386 \times 59 + 193 = 22967$ in., distance fallen in the last second. 193 in. $+ 22967$ in. $= 23160$ in.; $23160 \times 60 = 1389600$: 1389600 in. $\div 2 = 694800$ in. $= 57900$ ft., *Ans.*

Geometrical Progression.

Case I.

Art. 269.

(3.) $2^{12} = 4096$; $4096 \times 2 = 8192$, *Ans.*

(4.) $4^8 = 65536$; $262144 \div 65536 = 4$, *Ans.*

(5.) Ratio $= 3$; $3^9 = 19683$; $19683 \times 10 = 196830$, *Ans.*

Case II.

Art. 270.

(2.) $3^6 = 729$; $729 \times 10 = 7290$, last term. $7290 \times 3 = 21870$; $21870 - 10 = 21860$: $21860 \div 2 = 10930$, *Ans.*

(3.) $2^{11} = 2048$; $2048 \times 1 = 2048$, last term. $2048 \times 2 = 4096$; $4096 - 1 = 4095$, and $4095 \div 1 = 4095$, *Ans.*

(4.) $4^{11} \times 4194304$; $4194304 \times 1 = 4194304$, last term. $4194304 \times 4 = 16777216$; $16777216 - 1 = 16777215$; $4 - 1 = 3$; $16777215 \div 3 = 5592405$ ct. $=$ \$55924.05, *Ans.*

(5.) $.3 \times 10 = 3$; $10 - 1 = 9$; $3 \div 9 = \frac{1}{3}$, *Ans.*

(6.) Ratio $= 3$; $\frac{1}{3} \times 3 = 1$; $3 - 1 = 2$; $1 \div 2 = \frac{1}{2}$, *Ans.*

(7.) Ratio $= 2$; $\frac{1}{2} \times 2 = 1$; $2 - 1 = 1$; $1 \div 1 = 1$, *Ans.*